THE MICROECONOMICS OF
TECHNOLOGICAL SYSTEMS

The Microeconomics of Technological Systems

CRISTIANO ANTONELLI

OXFORD
UNIVERSITY PRESS

This book has been printed digitally and produced in a standard specification in order to ensure its continuing availability

OXFORD
UNIVERSITY PRESS

Great Clarendon Street, Oxford OX2 6DP
Oxford University Press is a department of the University of Oxford.
It furthers the University's objective of excellence in research, scholarship, and education by publishing worldwide in
Oxford New York
Auckland Cape Town Dar es Salaam Hong Kong Karachi
Kuala Lumpur Madrid Melbourne Mexico City Nairobi
New Delhi Shanghai Taipei Toronto
With offices in
Argentina Austria Brazil Chile Czech Republic France Greece
Guatemala Hungary Italy Japan South Korea Poland Portugal
Singapore Switzerland Thailand Turkey Ukraine Vietnam

Oxford is a registered trade mark of Oxford University Press
in the UK and in certain other countries

Published in the United States
by Oxford University Press Inc., New York

© Cristiano Antonelli, 2001

The moral rights of the author have been asserted

Database right Oxford University Press (maker)

Reprinted 2009

All rights reserved. No part of this publication may be reproduced, stored in a retrieval system, or transmitted, in any form or by any means, without the prior permission in writing of Oxford University Press, or as expressly permitted by law, or under terms agreed with the appropriate reprographics rights organization. Enquiries concerning reproduction outside the scope of the above should be sent to the Rights Department, Oxford University Press, at the address above

You must not circulate this book in any other binding or cover
And you must impose this same condition on any acquirer

ISBN 978-0-19-924553-6

Acknowledgements

Preliminary versions and parts of the book have been presented at the IRIS workshop of Artimino in September 1998; the conference 'Knowledge Spillovers and the Geography of Innovation: A Comparison of National Systems of Innovation' organized by the Centre de Recherche Économique of the University of Saint-Etienne, July 1999; the Seventeenth Annual Conference of the European Association for Research in Industrial Economics in Turin, September 1999; the Regional Conference of the International Telecommunications Society in Turin, September 1999; the Fortieth Annual Conference of the Società Italiana degli Economisti in Ancona, October 1999; the international workshop 'The Political Economy of Technology in Developing Countries', organized by INTECH, the Institute for New Technology of The United Nations University in Brighton, October 1999; the conference 'Collective Invention and European Policies' organized by the IMRI of the University of Paris IX Dauphine in Paris, October 1999. The book also includes material drawn from papers prepared respectively for the workshop 'Connaissances Organization et Activités Technologiques' of Aix-en-Provence, March 2000; the conference 'New Frontiers in the Economics of Innovation and New Technology' at the Accademia delle Scienze in Turin in May 2000; the conference 'Tecnologia e Società' at the Accademia dei Lincei in Rome, December 2000; the conference 'Toward a Learning Society: Innovation and Competence Building with Social Cohesion for Europe', organized with the sponsorship of the European Commission, Key-Action on Socio-Economic Research, 28–30 May 2000 in Lisbon, the workshop on the regional level of implementation of innovation and education and training policies, Brussels, November 2000; and the International Schumpeter Society Conferences in Vienna, June 1998 and Manchester, June 2000 as well as seminars at the IDEFI of the University of Nice and the SPRU of the University of Sussex in spring 2000, the Bank of Italy in Rome, October 2000, and the lectures at the European Summer School on Industrial Dynamics, in Cargese, September 2000. The book also draws from the lectures prepared for the European Chair at the University of Paris IX

Dauphine through the year 2000. The membership of the Board of Directors of Telecom Italia and of the Scientific Advisory Board of ENEA has been especially important and a rich source of learning opportunities. The book is partly based upon the systematic redrafting and reorganization of material previously published by *Regional Studies, Review of Economic Conditions of Italy*, the *Oxford Handbook of Economic Geography*, edited by G. L. Clark, M. Feldman, and M. S. Gertler (Oxford University Press, 2000), *The International Handbook on Telecommunications Economics*, edited by G. Madden and S. Savage (Edward Elgar, forthcoming), the book *Knowledge Spillovers and the Geography of Innovation: Institutions and Systems of Innovation*, edited by Maryann Feldman and Nadine Massard, (Dordrecht: Kluwer, forthcoming). This book draws from the large and detailed case-study evidence about the dynamics of technological knowledge in technological districts made available in the final report to the European Union for the research project 'Industrial Districts and Localised Technological Knowledge'. The support of the European Union TSER projects, 'Industrial Districts and Localised Technological Knowledge' and 'Collective Invention and European Policies' is acknowledged together with the funding and cooperation of the Programme 'Sistemi tecnologici, valutazione della ricerca e politiche per l'innovazione' of the Consiglio Nazionale delle Ricerche (CNR research grants nos. 96.03949.PS10, 97.05173.CT10, and 98.03795.PS10); the European Union Programme 'LEONARDO/EMET' and the Project 'Competenze, networks e dinamica industriale' of the MURST. The comments and suggestions of Giacomo Becattini, Jacques De Bandt, Aldo Enrietti, Maryann Feldman, Dominique Foray, Aldo Geuna, Jackie Kraffts, Roberto Marchionatti, Stan Metcalfe, Michel Quéré, Joel and Jacques Ravix, Ed Steinmueller, and Peter Swann, at various stages of the work, as well as the comments of many members of the TSER and CNR working groups, are gratefully acknowledged.

Contents

1. Introduction — 1
2. Irreversibility, Innovation, and Technological Complementarities — 13
3. The Economics of Localized Technological Knowledge: Learning Recombination and Increasing Returns — 44
4. Dynamic Efficiency Wages, Learning, and Innovation — 55
5. Localized Technological Knowledge as a Collective Activity — 69
6. The Economics of Technological Communication in Technological Districts — 81
7. Collective Knowledge and the Dynamics of Technological Clusters: The Case of Communication Technologies — 111
8. The Dynamics of Knowledge Internalization: The Case of Fiat in the Technological District of Turin in the Mechanical Engineering Cluster — 146
9. New Directions in the Corporate Production of Technological Knowledge — 181
10. Conclusions — 195

References — 212
Index — 227

To Anna,
With Whom Dreams Come True

1

Introduction

This book is dedicated to enquiring into the systemic and dynamic characteristic of the generation of technological knowledge and introduction of technological change. It contributes to the new debate on the economics of technological knowledge and innovation with a number of qualifications and specifications about the hypotheses of increasing returns in the production of knowledge and hence in the economy at large. It provides a general framework to understand the economic and institutional determinants of the regional and technological concentration of innovation activities. To do so the book elaborates an interpretative framework for understanding the long-term interactions between location in regional space, irreversibility, localized technological knowledge, and change and complementarity. The book builds upon the results of systematic explorations into the microdynamics of technological knowledge and technological change within a large number of European technological districts and the technological clusters centred upon new communication and mechanical engineering technologies.

This book is also an enquiry about two distinct and yet complementary meanings of the notion of localized technological knowledge and their relations with the notion of localized technological change. The first refers to the technical space and defines technological knowledge in terms of technological contiguity and complementarity that impinge upon common scientific and technological opportunities. The applications of localized technological knowledge into complementary and interdependent technological innovations constitute a technological cluster. The second makes reference to the regional space and defines the technological knowledge which is fed, within technological districts, by local interactions and communication routines in a limited regional space. The notion of localized technological change focuses the actual introduction of new technologies that are localized in proximity to previous ones in terms of factor intensity. The reference here is clearly the original notion of localized technological change introduced by Atkinson and Stiglitz (1969). In all cases localization is at the same time a key aspect in the definition and a causal factor for analysing the generation of new technological knowledge and the introduction of new technologies. The basic object of analysis of this book is in fact the interaction between the localization of technological knowledge in the technical space, within technological clusters, and its localization in the regional space, within technological districts, and how and why these two distinct aspects are in fact complementary dimensions of the systemic character of both technological knowledge and technological change and are conducive to the accelerated generation of new technological knowledge and introduction of localized technological innovations.

The uneven distribution of innovation activities, in terms of inputs, such as research and development expenditures and learning; outputs, such as patents, innovation counts, and total factor productivity growth rates; and effectiveness, in terms of the relationship between innovation inputs and outputs, across technologies and regions, is now well documented. The clustering of fast rates of accumulation of new technological knowledge and introduction of technological changes in well-defined technological districts and technological clusters has been the object of growing attention in recent economic analysis.

Much empirical evidence recently gathered in regional and innovation economics has revealed that innovation activity is strongly concentrated in regional space. Innovation activities cluster in a few regions with high levels of regional concentration, not only within countries but even at a global level. Worldwide regional concentration in innovation activities is higher than regional concentration in employment and output. Regional concentration in innovation activities is higher than firm concentration in the same innovation activities. This evidence is confirmed using a variety of indicators: patents statistics indicate that patenting agents cluster in a limited number of regions. Industrial analyses confirm that innovative activities in specific industries are localized in a few regions. Total factor productivity analysis suggests that a significant spread persists over time across regions. Research and development

Introduction 3

indicators confirm that a major share of research activities is also concentrated in a few regions worldwide (Clark et al., 2000). This evidence provides new support for the Marshallian notion of external economies (Marshall, 1920/1961). Specifically, it confirms the important role of the clustering of firms in industrial districts as a source of technological opportunities and the need to identify the regional system as the relevant unit of analysis.

Empirical evidence on the direction of technological change suggests that high levels of technological concentration characterize innovation activities. Innovations tend to cluster in a few technologies. Technological concentration is different from industrial concentration in that it focuses and measures the distribution of technological advances across technological activities. Industrial concentration focuses and measures the concentration of innovation activities across industries, defined in terms of activity of firms. Finally, technological concentration also differs from firm concentration which in turn measures the distribution of innovation activities across firms. Technological concentration appears to be the result of the clustering of innovative activities in a limited range of technological fields and it is fed by the focused innovation activities of a wide range of firms active in a variety of industries. Since the late 1960s innovation activities seem to cluster in new information and communication technologies. A significant increase of innovation activities in biotechnologies has been observed since the early 1980s (Archibugi and Pianta, 1992; Archibugi and Michie, 1997, 1998).

The notion of general-purpose technologies has recently been proposed to take into account the effects of sequential complementarity among new technologies around a core of new enabling technologies and analyse their effects in terms of growth. In this context, however, little attention has been focused on the characteristics of the creation of such sequential complementarities. Their main features in terms of wide applicability across a broad range of uses and scope for improvement and elaboration are assumed to be exogenous (Bresnahan and Trajtenberg, 1995; Helpman, 1998).

This evidence and the new theorizing about general-purpose technologies, however, provide new support for the Schumpeterian intuition about the gales of innovation (Schumpeter, 1934). According to Schumpeter, technological innovations appear in gales and cannot be analysed in isolation. Also from a technological viewpoint, a systemic approach seems necessary.

On the basis of this evidence, an attempt to elaborate a systemic approach to understanding the microdynamics of technological knowledge and technological change is called for. The intrinsic diachronic and synchronic indivisibility of technological knowledge and technological change, together with the pervasive role of technological complementarities, limits the scope of individualistic analyses and indicates an appreciative theorizing approach where the non-price interactions of innovators are valued (Metcalfe, 1995a, 1995b).

This direction seems all the more necessary since at the macroeconomic level, a new body of literature, the so-called new growth theory, has suggested an interesting framework to explain endogenous technological change in the context of equilibrium analysis. According to Romer (1986, 1990, 1994), technological change is the result of the accumulation of two quite distinct forms of technological knowledge: generic technological knowledge, germane to a variety of uses, and specific technological knowledge, embodied in products and as such having highly idiosyncratic features. The latter can be appropriated, at least to a large extent; the former maintains the typical features of the Arrovian public good. The production of specific knowledge takes advantage of the availability of generic knowledge.

The new theory of growth has induced a wide array of specific applications and interpretations at the macroeconomic level and yet can provide an interesting stimulus to the microeconomics of innovation. The basic hypothesis which is the object of an intense debate consists in the reintroduction of increasing returns. Increasing returns, however, are now assumed to take place in the production of knowledge and are combined with decreasing returns in applications and hence in the production of specific goods. As Scherer (1999) aptly puts it, the basic novelty in the new growth theory consists in the new theorizing about the interaction between the two distinct components of knowledge—the generic and the specific—and in the assumption that such interaction leads to increasing returns: 'The human capital is made more productive by interacting with the stock of knowledge which includes knowledge of all designs previously achieved along with the scientific knowledge published by academic researchers. The more knowledge there is, the more productive R & D efforts, using human capital, are' (Scherer, 1999: 35).

The evidence about the regional and technological concentration of innovation activities and the theoretical underpinnings of the new growth theory can be matched so as to provide a larger framework for the economic analysis of technological knowledge and technological change. The basic argument can be summarized as follows. The evidence about the regional and technological concentration of innovation activities can be interpreted as confirmation that relevant increasing returns are at work in the production of technological knowledge and technological change. The limited number of regions, worldwide, and technologies in which such increasing returns take place, however, suggests that a number of important complementary economic and institutional factors are necessary for increasing returns actually to occur.

The book explores the system microdynamics of technological knowledge and technological change at the regional and technological level. It identifies the context and process conditions that make increasing returns possible and actually viable. In so doing, it values the determinants and effects of the localized search for new technologies. The introduction of new technologies is viewed as the result of a combination of relevant irreversibilities of production factors, entropy in the factors and products markets, and access to local externalities. Technological districts and technological clusters are, in fact, explained within a theoretical framework which values the combined and complementary effects of local externalities, irreversibility, and endogenous technological change as necessary conditions for dynamic increasing returns to occur in the production of new technological knowledge and in the introduction of technological innovations. In so doing, the book builds upon the tradition of analysis paved by the contributions of Paul David (1975, 1985, 1987, 1993, 1994, 1997, 1998).

In this context, an interpretative framework for understanding the determinants and effects of technological complementarities is set out. This provides an interpretative context in which to qualify the communication conditions that are necessary for increasing returns to occur in the production of knowledge and which can hence explain the regional and technological concentration of innovation activities.[1]

Evidence of the high levels of regional and technological concentration of innovation activities can in fact be accommodated in a proper theoretical framework only when it is assumed that firms cluster in a few 'innovation commons', that is, in technological systems such as technological districts and technological clusters when and if the common resources, in terms of external stocks of complementary technological knowledge, made available by proximity and repeated interactions, are not given and limited, but can be created and accrued with increasing returns.

In the economics of knowledge, the notion of modularity and near-decomposability of complex systems introduced by Herbert Simon plays a major role. Knowledge can be assimilated to a complex system and as such can be viewed as 'nearly decomposable', that is, articulated in subsystems. In a near-decomposable complex system, 'intracomponent linkages are generally stronger than intercomponent linkages. This fact has the effect of separating the high-frequency dynamics of a hierarchy—involving the internal structure of the components—from the low-frequency dynamics— involving interaction among components' (Simon, 1962/1969: 217). Local externalities, as opposed to global externalities, matter when and because technological complementarities are available, but within modules. Secondly and most importantly, in this approach, actual efforts are necessary to assimilate external knowledge. In such a context, the role of proximity both in regional and technological space and of technological communication is appreciated. Technological externalities are available mainly locally, within technological districts and technological clusters, within modules of technological complementarities, and are the result of a complex web of communication channels which need careful examination.

---

[1] In many ways this book elaborates systematically upon the blending of the Schumpeterian and the neo-Marshallian literatures and it is complementary itself to the previous *The Microdynamics of Technological Change* (1999a) where the analysis of the implications of the hypothesis of a localized character of knowledge and localized technological change focuses the effects of the increase in efficiency in a limited technical space. Here the hypothesis of the localized character of technological knowledge and technological change is developed with special attention to the effects in terms of complementarity among products and processes, within technological systems articulated in technological clusters and technological districts.

Introduction

Irreversibility matters in explaining the basic inducement mechanism which pushes firms to innovate. Substantial irreversibility costs exist and reduce the capability of firms to cope with the changing conditions of product and factor markets. When irreversibility matters, the introduction of technological innovations becomes an important alternative to standard technical substitution and, more generally, firms are induced to change the shape of their isoquants rather than moving along the existing ones.

All changes in relative prices and desired output levels oblige firms to change their existing combination of superfixed and variable factors. Irreversibility, however, makes technical substitution impossible and pushes firms into out-of-equilibrium conditions. Such conditions can become a powerful inducement factor to try and introduce localized technological changes. To avoid technical inefficiency, induced innovations are introduced along the endowment line, defined in terms of the original amounts of superfixed production factors. Hence technological complementarities are the endogenous result of the local search for new technologies induced and biased by the endowments of irreversible production factors. These latter, in turn, can be both internal and external to firms, but internal to regional and technological clusters. In the former case, endogenous technological complementarities lead to economies of scope; in the latter to technological externalities.

Standard microeconomics assumes the adaptive behaviour of firms. Firms adapt quantities to prices and vice versa without any possibility of generating endogenous changes in their technologies. Adaptive responses, however, are often made difficult by the irreversibility of production factors and relevant sunk costs. In these circumstances, firms can react by introducing endogenous structural changes. Firms change their technology and modify their production conditions. Creative reactivity, as opposed to adaptation, is the underlying theme of our analysis. The limitations to adaptations are explored and the conditions for reactivity are then assessed. As a matter of fact, proximity in regional and technological space plays a major role in both reducing the scope for adaptation and favouring the conditions for reactivity.

Regions are, at the same time, a source of major hysteretic constraints and innovation opportunity. Location is a major determinant of long-term rigidity and irreversibility as well as a context for technological communication, a source of external knowledge and learning opportunities: hence an important factor in assessing the rate and direction of technological change. Regions are a major factor in making technological change hysteretic. The interaction of the dynamics of localized technological changes and communication processes explains the clustering of innovations in well-defined technological districts as well as the rate of introduction of technological changes.

The efficiency of innovation routines becomes a key issue in this context. Our basic argument here is that the outcome of changes in the relative prices of production factors and in the desired levels of output on production processes shaped by the pervasive role of superfixed production factors, is influenced to a large extent by the properties of circumscribed modules of technological complementarities and local innovation systems in terms of communication capabilities and learning opportunities.

The generation of technological knowledge is a process with a strong collective character when and if it is the result of an activity that combines complementary pieces of information and knowledge that are owned by a variety of parties and cannot be fully traded as such. With low transaction and communication costs, technological complementarities can fully deploy their effects in terms of local increasing returns and positive feedback, within modules, on firms, technological clusters, and technological districts.

In this context, the conditions and features of communication processes explain the clustering of innovations in well-defined regional spaces and technological clusters. Localization in technological districts characterized by multi-layer communications systems favours access to external knowledge, now viewed as an essential intermediary input in the generation of technological knowledge, and promotes the introduction of localized technological changes, leading to self-reinforcing mechanisms based upon localized increasing returns.

The access to collective knowledge and the opportunities for technological pooling provided by effective communication systems, within technological districts, favour the efficiency of innovation activities within firms and the eventual introduction of localized technological changes. In turn, the enhanced efficiency of firms' innovation activities and faster rates of introduction of innovations increase the amount of collective knowledge available in the region. A spiralling interaction, fuelled by the localized positive feedbacks between firms and within regions and clusters, can take place with significant effects in terms of dynamic increasing returns both at regional, industrial, and firm levels.

The role of learning in the generation of new technological knowledge and hence technological innovation is acknowledged in the economic literature. Learning, however, cannot be considered a windfall: specific incentives are necessary in order to direct the creative efforts of employees towards the accumulation of tacit and eventually codified knowledge and hence actually to contribute to the introduction of technological innovation. This empirical evidence encourages us to revisit the efficiency wages literature and suggests some rethinking upon the relations between wages, learning, and innovation within technological districts.

Technological clusters and technological convergence are the results of the emergence of pools of collective knowledge. The analysis of collective knowledge and localized technological change provides a definition of technological clusters and makes it possible to explore their main features in terms of a variety of specific forms of interdependencies such as: technological, knowledge, pecuniary, direct and indirect, supply and demand network complementarities. In turn, such complementarities may translate into externalities, that is, direct and indirect effects, not fully mediated by the price system, among firms, as well as into economies of scope, within the borders of firms.

Analysis of knowledge and technological complementarities makes it possible to contribute to the theory of the firm, to assess the economic role of regional space, and to elaborate a tentative analysis of the sequential origins and effects of the gales of technological innovations within clusters.

Understanding of the main features of the new technological cluster emerging around communication technologies provides strong evidence about the key role of industrial dynamics in feeding the introduction of complementary technological innovations that draw from common pools of collective knowledge. The structure of this technological cluster appears to be the result of the continual entry of new firms and the horizontal and vertical diversification of incumbents in complementary industries aimed at valorizing endogenous economies of scope.

Analysis of the process of technological knowledge creation and the introduction of technological changes by Fiat, the leading

Italian manufacturing company, in the period since its foundation in 1900 to 1971 provides alternative and yet complementary evidence about the key role of endogenous economies of scope, technological districts, and technological clusters. This case study shows the key role of local accumulation of competencies in the early mechanical engineering industry in the technological district centred upon Turin, the progressive internalization of external competence by the company, and the key role of the creation of new internal complementarities and economies of scope. In this case, the growth of the corporation appears to be driven by a systematic process of internalization of external knowledge available, both in the technological district of Turin and in the module of technological complementarities centred upon the mechanical engineering technological cluster.

The new understanding of the relations between the system dynamics of collective knowledge in technological districts and technological clusters provides useful insights into the analysis of recent changes in the organization of technological knowledge production within large corporations. The relations between the dynamics of collective knowledge and industrial dynamics in terms of the conduct and performances of firms, both within technological districts and technological clusters, are also assessed with some implications for long-term industrial policy. While in the innovation policy tradition, public intervention is deemed necessary in order to cope with the intrinsic market failures associated with the economic production of knowledge, our argument about localized increasing returns in the production of knowledge in specific contexts of action, regionally and technologically defined, calls for a highly selective innovation policy, able to focus on a limited number of regions and technologies, and hence oriented towards the full display of the advantages of increasing returns. Such an innovation policy may have important market-enhancing effects, paving the way to the eventual creation of proper markets for new products.

In this context, the traditional analysis about the tragedy of commons can be reversed and the positive effects of the commons of technological knowledge, embedded in well-defined regions, can be appreciated. The notion of collective knowledge, viewed as the result of a dynamic accumulation process characterized by complementarity between the research and learning activities of a myriad of co-localized agents, makes it possible to understand the emergence of new technological clusters and plays a key role in assessing the innovation capabilities of innovation systems.

This approach makes it possible to lay down the basic elements of a system dynamics analysis of technological and economic change. This approach in fact leads to the identification and appreciation of the systemic elements which constrain and stimulate the innovative activity of firms, beyond the classical price–quantity adjustments. Synchronic and dynamic complementarities exert an array of direct and indirect effects on the behaviour and conduct of firms which are able to change their production functions and their technology. The price system reflects such effects only to a limited extent and only when technological change is considered an exogenous predictable event.

In our approach, firms are constrained by sunk costs and irreversibilities and are exposed to innovations that change their product and factor markets and hence affect their current and expected performances and profitability. Firms, however, can do more than adjusting prices to quantities and vice versa. They can innovate, generate new technological knowledge, and introduce new technologies. The search for new technologies is local, in the technical space defined by irreversible production factors. The larger are the systemic advantages provided by technological clusters and technological districts, based upon collective knowledge and related knowledge complementarities and externalities, the larger is the likelihood that they will be able to meet the new market conditions with the introduction of successful innovations. These latter, in turn, will change the market conditions of other firms which, in turn, may be able to react in a creative and innovative way. The foundations of an ongoing process of localized technological and economic change are put in place. In such a process, the external advantages provided by the communication conditions and hence by access to the collective knowledge available within modules of technological knowledge and technological systems play a key role.

The remainder of the book is organized as follows. Chapter 2 provides a simple model that shows how changes in demand and factor prices together with superfixed production factors induce the endogenous introduction of localized technological changes, the selection of new technologies according to their complementarity with existing technologies, and the implementation of research strategies. This chapter also provides a simple model for understanding how this process leads to new endogenous technological complementarities. In turn, technological complementarities can lead to economies of scope or technological externalities according to the internal or external character of irreversible production factors. Chapter 3 recalls the basic theory of localized technological knowledge. Chapter 4 focuses on the role of learning and explores the incentive system for firms to stimulate the active participation of workers in the production of knowledge. Chapter 5 provides a definition of technological knowledge as a collective good. In turn, Chapter 6 shows how the accumulation of collective knowledge and the economics of communication processes explain the clustering of innovations in well-defined regional spaces or technological districts. Chapter 7 explores the effects of collective knowledge in terms of the emergence of technological clusters. Chapter 8 provides evidence on the growth of a major corporation as a process of systematic accumulation of technological knowledge, based on internal learning processes and the internalization of technological opportunities and collective knowledge available in a technological district and a technological cluster. Chapter 9 elaborates the implications of the analysis for understanding the evolution of the organization of the technological knowledge production within firms. The implications for economic analysis and the relevant policy issues are considered in Chapter 10.

2

Irreversibility, Innovation, and Technological Complementarities

Superfixed production factors are a pervasive condition which economic theory is more and more taking into account. Technical interrelatedness, system bottlenecks, long-term obsolescence, limited span of application of intangible assets, location in well-defined regions are all factors that reduce the capability of firms to adjust, even in the long term, significant chunks of their stock of intangible and tangible capital to the changing conditions of both products and factors markets. In these conditions, all actual and expected changes in relative prices and desired output levels are likely to induce a significant decline in both output and price efficiency: firms are unable to produce with the 'correct' combination of flexible and fixed production factors. Standard textbook firms should accept their decline or the impossibility of taking advantage of new emerging market conditions for their products.

When irreversibility makes it impossible to match changes in product and factor markets with the full flexibility of production factors, however, firms can do more: they can try and change their technology so as to be able to increase locally total factor productivity and hence operate with the irreversible portions of their capital stocks and make productive use of this. In this approach, technological change is not only endogenous but also flexible. The introduction of new technologies is at the same time induced, directed, and selected according to the new technologies' compatibility and complementarity with the firm's endowment of irreversible production factors.

This chapter elaborates an interpretative framework for understanding the long-term interactions between irreversibility and localized technological change and their consequences in terms of endogenous growth and technological complementarities. Section 1 provides the basic argument for the distinction between adaptation and innovation. In Section 2, the role of irreversibility in economic analysis is briefly reconsidered. Section 3 presents in a simple model how changes in demand and factor prices together with superfixed production factors can induce the introduction of endogenous localized technological changes. Section 4 articulates the role of irreversibility as a selecting mechanism in a context of technological variety and introduces formally the notion of endogenous technological complementarities. In turn this Section shows that, according to either the external or internal character of irreversible production factors, technological externalities lead to economies of scope or technological externalities. Section 5 outlines the consequences of the analysis in terms of an endogenous and self-propelling growth process which takes place in out-of-equilibrium conditions. The implications for economic analysis are briefly summarized in the conclusions.

1. ADAPTATION AND INNOVATION

Standard microeconomics assumes the adaptive behaviour of firms. In the theory of production, agents adapt the technical combination of inputs to the relative price of production factors within a limited scope of substitution defined by a given production function. While the techniques can be changed, according to changes in the relative costs of production factors, the technology, defined in terms of general levels of total factor productivity and the marginal rates of substitution, is not under the intentional control of the firm as it is exogenous. In the market-place, the firm adapts quantities to prices and vice versa without any possibility of generating endogenous changes in their product technologies.

Adaptive responses, however, are often made difficult by the irreversibility of production factors and relevant sunk costs. Firms cannot adjust their factor intensities to the proper levels and cannot change the given levels of fixed production factors with respect to changes in desired levels of output. Both in product and factor markets disequilibrium conditions arise: the demand for production factors is altered by irreversibility as well as supply in the product markets. Product and factor prices are also, as a consequence, altered and kept away from correct levels. Product prices are below the correct levels and the demand for variable inputs is larger than it should be, with the consequence that prices for such production factors are larger than in equilibrium. In these circumstances, the whole system and the firms are out of equilibrium. The larger is the share of irreversible production factors in total, the further away is the system from standard equilibrium conditions, with adverse effects in terms of welfare losses. More specifically, firms characterized by irreversible production factors experience a loss of profitability and a general decline in performance. Both price and product inefficiencies arise.

Emergent technical inefficiency becomes all the more cogent in competitive markets with high levels of technical variety: firms with larger shares, of total costs, of superfixed production factors are soon exposed not only to a decrease of technical efficiency but also to an actual decline of market shares and profits. Such firms can survive in the market-place, although efficiency is substandard, because of the long-lasting economic life of irreversible capital goods. Sunk costs have already been paid for and firms experience losses which are for the most part virtual.

While standard microeconomics assumes that firms have only adaptive behaviour, the blending of the Schumpeterian and neo-Marshallian approaches suggests that firms can do more: they react. Reaction implies a creative capability and makes innovation possible. Reactivity is necessary when adaptation is made difficult by the context and specifically by irreversibility. Firms which experience serious limitations to their intended growth and adaptation, because of irreversibility, can confront emerging inefficiency by changing their technology.[1]

The search for new technological knowledge is triggered by the decline in actual and expected performance. Such an emergency can be accommodated by established firms with the endogenous generation of localized knowledge and the eventual introduction of localized technological changes that are compatible and complementary with superfixed production factors.[2]

---

[1] See Downie (1958) on the relationship between poor performances and the inducement to innovate. See also Finch (2000).
[2] See Herbert Simon (1958/1982): 'A direct and unsophisticated application of the theory developed in the preceding paragraphs would suggest that we should search for the initiation of innovation largely in the environment of the organization— particularly in adverse changes in the environment that threaten a previously existing

In our approach, additional technological knowledge is required in order to feed the creative reaction which can lead to the introduction of technological (and organizational) innovations and overcome the emerging losses stemming from the limitations to sheer adaptation, caused by irreversibility.

When and if localized technological knowledge has the features of collective goods within technological districts and technological clusters, with all the advantages of increasing returns in the generation of technological knowledge and in the introduction of technological innovations, it is clear that the larger the effects of increasing returns, the larger the likelihood that firms can react with the introduction of important innovations that make possible significant increases in total factor productivity.

Innovative reaction takes place in a context characterized by out-of-equilibrium conditions with both product and production prices which are, at the system level, away from equilibrium. Agents at large also have problems formulating rational expectations about the final exit from the market-place of firms with irreversible production factors. Creative reactivity to such market turbulence is clearly induced by a context of systematic market failures. Firms with irreversible production factors are exposed at the forefront of such turbulence. Rather than waiting for the eventual exit and/or mismatch between expected levels of output and price efficiency, firms can try and generate new technological knowledge and hence introduce new technologies under the constraint of the characteristics of the portions of existing and irreversible capital stocks.

Irreversibility matters not only with respect to history but also with respect to expectations. Irreversibility matters with respect to history when past investments exert long-lasting influence on today's decision-making. Irreversibility matters with respect to expectations when incumbents consider the constraints imposed by the future irreversibility of new investments on their growth path. In the former case, agents 'discover' the irreversibility of portions

---

level of achievement. Two kinds of adverse conditions come to mind at once: a downturn or anticipated downturn in the business cycle; innovations by competitors that improve their market position' (p. 395). 'As a result, the adaptation of organizations to new circumstances is often achieved by local change, rather than by evolutionary innovations that alter the organizational programs simultaneously over a wide area' (p. 398).

of their capital stock. In the latter, they are aware of the superfixed character of the new investments. In both cases, we assume that firms can react by making endogenous structural changes so as to introduce new processes and new products which take into account the endowment, either existing or expected, of superfixed production factors (Krugman, 1991a).

Firms can change their technology and modify their production conditions by introducing new products and new processes. Both product and process innovations will be the result of an innovation process both induced and constrained by the endowment of irreversible inputs and hence by the incentive to make productive use of them. New process technologies will be introduced so as to reuse existing capital goods. New products will be introduced so as to take advantage of latent technological complementarities. Actually, technological complementarities and related technological externalities and economies of scope are the endogenous result of the innovation strategy of firms which are at the same time induced to innovate and constrained in the direction of their innovation strategy by the features of the irreversible portion of both tangible and intangible capital.

Irreversibility at the same time becomes a powerful inducement mechanism to introduce new technologies, a focusing device to direct research strategies towards specific factor intensities and product specifications, and last but not least a powerful sorting device to select and rank new technologies, according to their compatibility and interoperability with the existing stocks of irreversible production factors.

This localized search for new technologies, the consequent creation of endogenous complementarity, and its relationship with irreversible production factors can be assimilated to the biological process named exaptation.[3] In evolutionary biology and psychology, exaptation denotes the creative process by means of which species are able to make new use of existing and irreversible genotypes. Exaptation, in other words, can be considered an interface between the Darwinistic process of genetic selection and the Lamarkian dynamics of creative learning. In exaptation, phenotypes can interact with genotypes.

---

[3] See Rizzello (1999); Gould and Vrba (1982); Gould (1991).

Creative reactivity, as opposed to adaptation, is the underlying theme of our analysis. The limitations to adaptation are explored and then the conditions for reactivity are assessed. As a matter of fact, regions play a major role in both reducing the scope for adaptation and favouring the conditions for reactivity. In turn, reactivity is easier in a regional context—the technological district—which is conducive to the accumulation of technological knowledge and the eventual introduction of technological innovation.

2. IRREVERSIBILITY AND TECHNOLOGICAL CHANGE

Fixed production factors have always attracted attention as far as their effects on the production process and the distribution of revenue are concerned. Alfred Marshall elaborated the notion of quasi-rents to explain the discrepancies between expectations and actual circumstances in the management of the production process with respect to factor productivity and returns from investments:

> But if he invests in land, or in a durable building or machine, the return which he gets from his investment may vary greatly from his expectations. It will be governed by the market for his products, which may change its character largely through new inventions, changes in fashion, etc. during the life of a machine, to say nothing of the perpetual life of land. The incomes which he thus may derive from investments in land and machinery differ from his individual point of view mainly in the longer life of land. But in regard to production in general, a dominant difference between the two lies in the fact that the supply of land is fixed (though in a new country, the supply of land utilised in man's service may be increased); while the supply of machines can be increased without limit. At this difference reacts the individual producer. (Marshall, 1920/1961: 341)

An array of detailed empirical analyses, especially in industrial economics, has highlighted the key role of superfixed production factors (Antonelli, 1999a). Superfixed production factors are long-lasting, tangible, and intangible assets which can be changed to a limited extent and only in the very long term. Such assets, once installed, can be replaced only with huge costs and a long time span. Recent advances in the theory of investment are now dealing with such empirical evidence (Dixit, 1992). The analysis of sunk costs has made an important contribution to the theory of the firm and markets (Sutton, 1991). This chapter tries to model the effects of superfixed production factors on the rate and direction of technological change. Besides delayed economic and technical obsolescence, technical interrelatedness, which takes the form of major technical constraints among phases of the production process and between capital, intermediary goods, and skills; subsystem bottlenecks and complementary assets, dedicated and idiosyncratic competence are all factors which reduce the range of technical choices facing firms and keep them in a limited region of techniques within the existing map of isoquants. Sunk costs are especially relevant when the discrepancy between the purchasing costs of capital goods and resale prices in secondary markets is high: this is the case with most intangible assets. The idiosyncratic characteristics of each firm's production process add on to make evident the superfixed character of a significant portion of their production factors. Finally and most importantly, location in a well-defined regional space is a major factor of rigidity. Location roots firms in a variety of ways: plants and buildings are often difficult to change and expand; user–producer relations in intermediary markets have a strong regional aspect, as do internal and external labour markets. Regions are a major factor of irreversibility.

Most industries are characterized by interrelatedness and irreversibility of their capital stock. In many manufacturing industries, such as chemicals, steel, intermediary inputs, electronics, energy, production processes can be changed only with major problems and new plants can actually start only five to eight years after their start-up. In most network industries, such as telecommunications, transportation, electrical power, gas and water, railways, complementarity and compatibility among production units are a major constraint and changes can be introduced either when compatibility and interoperability between different vintages of capital goods are strictly enforced or when a drastic write-off of the whole system is undertaken.

It seems important to stress that because of irreversibility and interrelatedness of large portions of fixed capital, in many industries investment decisions and related technical choices are mainly based upon the expectations of firms with respect to the future conditions of factor and product markets. Equilibrium conditions are in these circumstances mainly expected equilibrium conditions rather than actual ones. The discrepancy between expected equilibrium and actual market conditions plays a major role when production factors are characterized by irreversibility and interrrelatedness.

When superfixed production factors are relevant, in that they constitute a major part of total production factors, all adjustments by firms to changing conditions in the business environment are subject to significant constraints. Changes in the production mix and size of output expose firms to relevant price and output 'Farrell' inefficiency, with a sharp decline in the quasi-rents associated with the efficient use of superfixed production factors.

Standard microeconomic analysis fully recognizes the importance of these factors: it seems relevant to recall that much production and costs theory, as well as all market theory, is based, as a matter of fact, almost exclusively in the short term, that is, the period of time during which fixed production factors cannot be changed. The short term is the single justification for U-shaped average costs and the related positive slopes of marginal cost and supply schedules. Short-term and long-term cost theories are reconciled only when competitive equilibrium applies: in equilibrium firms produce at minimum average costs which are consistent with their current endowment of fixed production factors (Stigler, 1939).

When out-of-equilibrium conditions are considered, however, a clear divergence emerges between *ex ante* and *ex post*: such divergence is all the stronger and analytically relevant when the role of long-term expectations is considered.[4] Short-term and long-term cost theories can be consistent not only in actual equilibrium conditions but in perspective equilibrium as well, when investment

---

[4] Stigler (1966: 130) falls short of introducing the notion that the design (the technology?) of a plant is an endogenoeus variable that the firm tries to take into account when dealing with fluctuating output 'We shall argue that the productive service which we arbitrarily hold constant in order to exhibit diminishing marginal returns is often actually fixed for the entrepreneur in the short run. Then he cannot make any magical transformation of the constant productive factor—it requires time to wear out such factors (if they are durable) or to rebuild them. Since the firm will nevertheless usually have a fluctuating output even in the short run, the entrepreneurs will seek to have a flexible productive system—one which operates with tolerable efficiency over a considerable range of outputs. This flexibility can usually be achieved at a [cost]: for example it is possible to design an oil refinery so it can vary substantially the proportions into which gasoline, fuel oil, and other products are obtained from given crude oil. In terms of our diagram, the flexible plant will have a lower output at X because, if versatility is expensive, a larger quantity of constant factor is needed, but the marginal product will not fall so rapidly when the variable productive service is increased.'

decisions have to be made with a long historic time horizon because of the low levels of physical and economic obsolescence of the capital goods acquired (Amendola and Gaffard, 1988).

With respect to such a tradition of analysis, an important step forward can be made when the possibility that technology can change is considered. Specifically, this opens two avenues of analysis: diffusion theory and innovation theory.

The former analyses conditions for firms which have to decide whether to adopt new technologies that have been introduced by third parties or exogenously in the system, according to their endowment of superfixed production factors. This departure from short-term production analysis and the related implications for cost theory finds an important precedent in the path-breaking analysis of Salter (1966) who has shown how (super)fixed production factors delay the adoption of new capital goods.[5] The durability of existing capital goods in fact, in Salter's model, delays the purchase of innovated capital goods until the variable costs associated with existing machinery are lower than the total average costs, including fixed costs, of the innovated machinery. Hence, new innovated capital goods are not adopted because of the effects of existing capital goods. Diffusion, that is, the non-instantaneous adoption of superior capital goods and the time distribution of adoptions among agents, is explained by the heterogeneous vintages of capital stocks in place among agents and relatedly by their rates of growth of demand. The larger is demand growth, the faster are diffusion rates, for given vintages.

---

[5] See Salter (1966: 4–5): 'The crux of the difficulty lies in the inability of static equilibrium concepts to analyse continuous processes over time. Let me illustrate this with an example drawn from the theory of production. In discussing the productivity of a factor of production we must obviously take note of its price, and the theory of production provides a convenient apparatus for analysing the effect of changes in relative factor prices on the productivity of different factors. Consider the case where entrepreneurs wish to employ more highly mechanised techniques because of a change in relative factor prices. The change-over is a slow process, for as Professor Hicks has said "... an entrepreneur by investing in fixed capital equipment gives hostages to fortune. So long as the plant is in existence, the possibility of economising by changing the method or scale of production is small; but as the plant comes to be renewed it will be in his interest to make a radical change" (*The Theory of Wages*, London, 1932, p. 183). This is the difficulty: certain adjustments to changing conditions take long periods of time to work themselves out, particularly when capital equipment is involved'.

The latter can explore the conditions into which firms, with existing capital goods, in out-of-equilibrium conditions, can generate technological knowledge and introduce technological changes: technological change is now endogenous. Firms constrained by superfixed production factors in the map of existing techniques can consider the possibility of introducing new technologies that make it possible to minimize the costs of irreversibility.

3. THE INDUCEMENT OF LOCALIZED TECHNOLOGICAL CHANGE: A GEOMETRIC EXPOSITION

In this context, the endogenous introduction of localized technological changes can be considered the result of the optimizing behaviour of myopic agents constrained by their endowment of fixed production factors.[6] So far this model provides an alternative and yet complementary analysis to the dynamics of localized technological changes induced by the interplay between switching costs and learning capabilities (Atkinson and Stiglitz, 1969; David, 1975; Antonelli, 1995, 1999a).

Firms which are exposed to changes in the relative prices of factor costs and/or in the levels of their demand may incur significant declines in technical efficiency and a fall in the quasi-rents associated with the endowment of superfixed production factors. In these conditions, firms are induced to try and find a solution in the form of a new technology. In this case they will incur innovation costs, that is, the costs of implementing their tacit knowledge and actually changing their technology.

More precisely, we consider two cases: when relative prices of production factors change and when demand levels increase. Let us consider a firm in equilibrium with a given level of superfixed production factors and a ratio of wages W to rental costs R at point A. After a compensating change in relative prices creating a new level of wages W' and rental costs R', the firm of the standard microeconomics textbook would choose the new technique B where the new

---

[6] Far-sighted firms might be able to spot the advantages provided by new technologies without any specific inducement mechanism. The analysis on the limitations to Olympic rationality elaborated by Herbert Simon seems to apply all the more to future technologies and related input and factor markets conditions.

marginal rate of substitution equals the slope of the new relative prices (see Fig. 2.1). A similar process takes place when the firm is exposed to increases in the levels of their demand. When demand increases, textbook firms should increase the levels of inputs, with a given technique, in order to expand output to the new desired level. Now that firm would reach point B on the new isoquant placed further to the right on the same map. The combination of both changes increases the impact of the situation.

Such solutions, however, imply that the levels of superfixed production factors will change: typically, this is a very long-term solution which cannot be considered. It is now clear that our context of analysis involves an extension of the time horizon of the traditional short-term cost and production analysis.

The firm with superfixed production factors cannot do any better than selecting the technique A, defined by the intersection between the isoquant and the endowment axis. Alternatively, the incumbent can select the technique defined by the intersection between the new isocost and the endowment axis, but for a given map of isoquants and hence for a given technology, the new solution implies a clear output inefficiency in terms of lower levels of output. In both cases, the solution reveals a discrepancy between the 'correct' isocost and isoquant, which of course are tangential in equilibrium. In these conditions, it is clear that a firm exposed to significant changes either in the demand for its product or in the relative price of its production factors is bound to experience either a decline in price efficiency, specifically in terms of the reduction in quasi-rents associated with superfixed production factors, or an emerging output inefficiency (Farrell, 1957). More directly, bearing in mind the Marshallian tradition of analysis, we can define the effects of such situations as quasi-losses.

In sum, it should be clear that all changes in the business environment, with respect to expected levels of output and factor costs, generate a reduction in general efficiency which is directly related to the distance on the isoquant map between the new desired equilibrium point and the one actually possible, clearly defined, for a given output, by the intersection between the isoquant and the endowment of superfixed production factors.

In these circumstances, the implementation of tacit knowledge, the generation of localized technological knowledge, and the introduction of new technologies that make possible the re-establishment of the efficiency conditions become a viable alternative to quasi-losses.

This inducement is all the stronger, the more competitive the market-place, the lower the barriers to entry, and the larger the variety of firms in terms of superfixed production factors endowment. In a highly competitive market-place, quasi-losses directly affect the market-place for incumbents which face new competitors that are unconstrained by superfixed production factors endowment and hence more efficient: quasi-losses may soon become losses. The same is true when incumbents differ in terms of the stock of superfixed production factors: the larger it is, the heavier the quasi-losses and hence the risk of facing not only a decline in profitability but also the emergence of actual losses. Hence, the larger the variety of firms and the more competitive the market-place, the more likely is the introduction of localized technological changes which enable incumbents to restore their output and price efficiency.

Finally, and for the same reason, the larger the entropy in factor markets and the larger their turbulence, the stronger is likely to be the inducement to generate new localized technological knowledge and introduce localized technological changes.

Bearing all this in mind, we can now turn to a brief formal exposition of the model.

We can write the quasi-loss function associated with both the changes in the relative prices of inputs and in the demand for incumbents characterized by relevant superfixed production factors, as follows:

$$QL = f(dAX) \tag{2.1}$$

where QL are the quasi-losses measured in terms of the technical distance between A, on the isoquant line and any new solution X on the new isocost for the new level of output on the endowment axis.

We can now turn our attention to the role of technological knowledge and technological change. We assume, in fact, that because of the strong role of learning in acquiring the tacit character of localized knowledge that is necessary to innovate and because of the complementarity between tacit knowledge and R & D activities, the search for new technologies is especially productive along the endowment of superfixed production factors expressed by the endowment axis AA'.

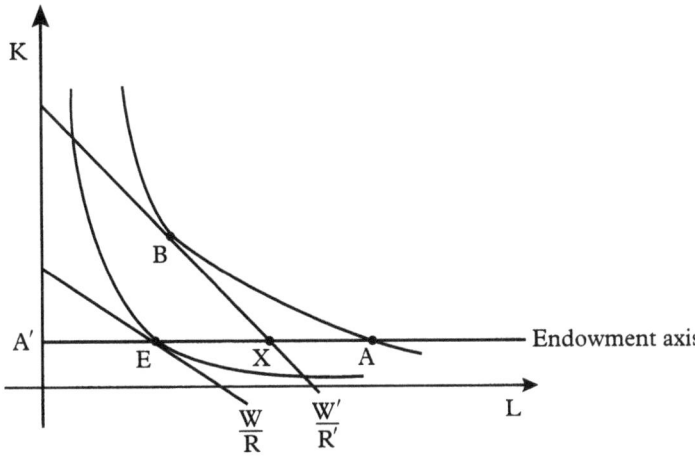

Figure 2.1. *Irreversibility and localized technological change.*

In fact, all new technologies that reshape the isoquant map along the endowment axis AA′ enable the productive factors to be used rationally so as to restore the general price efficiency of the firm and lead to an overall increase in efficiency and eventually in total factor productivity.

The introduction of new and better technology is, however, costly. In order to capitalize on the tacit knowledge acquired by means of the learning that has been going on in the techniques being used, firms have to invest in formal R & D activities. A systematic search for available external knowledge is also necessary and relevant communication costs are associated with this. This research process can stop when the new technology is such that the firm reaches point X. The firm is, in fact, again in equilibrium because the marginal rate of substitution again equals the slope of the new relative prices. Further movements along the AA′ axis beyond X, however, are welcomed. They are actually likely to generate an increase in total factor productivity in absolute terms.[7]

---

[7] An important reference in this context is of course the modelling tradition introduced by Nordhaus (1969) and implemented by Dasgupta and Stiglitz (1980) in the context of the theory of oligopolistic markets. From this viewpoint our approach can be considered as an appraisal of the Nordhaus model, strictly implemented within the boundaries of a 'short-term' theory of production.

The cost of the innovation activities necessary to move the isoquant along the endowment axis AA' towards point X, and further left, is a function of the leftward distance from the original equilibrium technique:

$$CTI = g(dXA) \qquad (2.2)$$

where CTI represents the cost of the resources dedicated to implementing learning procedures, building technological communication channels with other firms and with other research institutions, operating R & D laboratories, and broadly of all the activities directed towards the introduction of the technological changes that are necessary to reshape the isoquant so as to move along the line AA' and restore the tangency between the short-term solution A and the point X. [8]

A firm that chooses to continue with technique A incurs a decline in general price and output efficiency but avoids all innovation costs. Conversely, a firm that chooses a new technology which makes technique X efficient, that is, a tangent between isocosts and isoquants, avoids a decline in efficiency, but incurs the substantial innovation costs which are necessary to discover the new technology which will enable the firm to produce as much as on the old isoquant but next to or actually beyond the intersection between the new isocost and the endowment line.

We are now in a position to portray the decision-making process of the firm with the standard tools of profit maximization. The profit equation for the firm reads as follows:

$$P = R(dAX) - CT(dAX) \qquad (2.3)$$

where R stands for the gross revenue from adjusting to the new factor prices, measured in terms of the reduction in production costs, with respect to the distance between the technique X and the technique A, made possible by the introduction of technological changes that reduce price and output inefficiency. In other words, the revenue from changing the technology along the line AA' consists of

---

[8] The 'necessary' assumption that $g'' > 0$ seems plausible.

the reduction in the total costs in A.[9] CT(dAX) are the innovation costs and can be measured in terms of the distance along the endowment line between A and X.

Standard maximization of the profit equation enables the identification of the 'correct' amount of research expenditures a firm can bear. Our geometric approach makes it possible to relate the amount of innovation selected directly to the technical inefficiency arising from superfixed production factors, so as to establish a trade-off between technical inefficiency and technological innovation. Maximization here identifies the 'best' distance on the endowment line a firm can travel by means of the introduction of localized technological changes, induced by the twin constraints of a production process characterized by significant superfixed inputs and changes both in demand and relative prices in the business environment.

With low innovation costs, the incumbent may be induced to cover a long distance along the endowment axis and actually go beyond the intersection between the new isocost and the endowment line so as to introduce significant technological changes which enable a substantial increase in total factor productivity.

With high innovation costs, incumbents will be induced to make only incremental innovations so as to reduce the distance on the endowment axis between A and the intersection between the isocost and the endowment lines. In these circumstances, incumbents can only reduce quasi-losses and come closer to best practice.

The introduction of localized technological changes along the endowment axis will make it possible for firms that cope with the increase in demand levels to adjust the ratio of marginal productivities, actually increasing the usage intensity of the superfixed production factor. The direction of technological change will be shaped by the endowment of superfixed production factors. When the external shock consists of growth in product demand, the new technology will, in fact, exhibit a larger output elasticity for the production factors that happen to be 'flexible'. The rate of technological change, in turn, will be affected by the levels of entropy of the business environment and the share of superfixed production

---

[9] The 'necessary' assumption that $dR/dAX = 0$ seems plausible for the reduction in costs. The latter can be considered as the revenue of innovation activities which make possible the introduction of localized technological innovations along the fixed capital endowment line. This can be modelled as linear in the distance from the intersection of the isoquant expressing the new desired output level.

factors in total costs. The larger are both the former and the latter, the larger is the inducement to rely upon technological change in order to cope with new price and market conditions.

Similar results can be obtained when variations in the expected equilibrium position are induced by changes in relative prices of production factors. All changes in relative prices of production factors would oblige firms to modify the ratio of reversible to irreversible production factors. Once more the introduction of technological changes, now induced by factor cost changes, would make it possible to reduce price inefficiency provided firms are allowed to restore an equilibrium position which assumes the irreversible nature of fixed production factors.

In general, we see that the process is likely to reduce the quasi-losses associated with the 'wrong' mix of production factors engendered by the presence of superfixed features in the production mix. Specifically, moreover, it is clear that with efficient innovation activities and high levels of technological and learning opportunities firms can actually go beyond the 'equilibrium' point which simply restores the desired 'Farrell' efficiency conditions and find new technological conditions that actually increase total factor productivity. For given levels of endowment of superfixed production factors and entropy in factor and product markets, an economic system is likely to increase its total factor productivity levels the larger the efficiency of innovation activities and the more conducive the local innovation system to the generation of new technological knowledge (Nelson, 1987).

This approach to understanding the dynamics of irreversibility and the introduction of technological change has many implications:

(1) technological change induced by irreversibility is localized in that it takes place in a well-defined technical space and consists in the introduction of new technologies within a limited range of input proportions;

(2) the direction of technological change is clearly shaped by the endowment of irreversible production factors and their share with respect to reversible factors. Technological change, introduced by incumbents, is hysteretic in that superfixed production factors define the set of techniques with respect to which firms have an incentive to introduce technological innovations;

(3) the new technology will be more and more fixed-input-intensive when the discrepancies between expectations and actual factor and market conditions push firms to search for new technological solutions to the left of the expected equilibrium position. On the other hand, when actual market conditions induce firms to operate to the right of the expected equilibrium position, the new technologies are likely to be flexible-input-intensive. This bifurcation in the outcome of the localized process of introducing new technologies is all the more interesting when it is combined with a macroeconomic policy analysis: all (unexpected) stimulations to aggregate demand are likely to induce the introduction of new flexible-input-intensive technological change with a significant increase in derived demand for them (employment), while all unexpected reductions of demand levels may engender a fall in the demand for flexible inputs (employment). By the same token, it should be also clear that all unexpected reductions in the price of either production factor is likely to encourage the introduction of new flexible-input-intensive technologies and, on the contrary, all unexpected increases in the cost of either production factor is likely to induce the introduction of new fixed-input-intensive technologies;

(4) the rate of technological change depends now on both the endowment of irreversible production factors and the characteristics of the technological environment in which each firm operates: hence the larger is access to technological communication and the greater are the opportunities to take advantage of knowledge externalities, the higher is likely to be the rate of introduction of new technologies for firms facing changes either in demand or relative prices. In countries with efficient innovation systems, actual total factor productivity levels are likely to increase beyond current efficiency frontiers, while in countries with low connectivity levels and poor technological and learning opportunities, technological change will simply help firms to achieve expected productivity levels, reducing quasi-losses, but without actual movements of their efficiency frontiers;

(5) regions with large endowments of capital and consequently lower relative capital rental costs are more exposed than labour-abundant countries to the effects of irreversibility and hence are likely to experience stronger inducements to generate new localized technological knowledge and introduce localized technological changes;

(6) the larger the entropy of the economic systems and the wider the discrepancies between expectations and actual market conditions, both for products and factors, and the higher is industrial entropy, the larger are likely to be the rates of introduction of new technologies, when the technological environment is conducive to facilitating the actual search for and generation of new technologies;

(7) the search for new technologies and the direction of research and development activities are very much shaped by the constraints of the technical idiosyncratic character of existing technologies: new technologies, in order to be successfully introduced, must be compatible, fungible, and complementary with existing ones. Complementarity, compatibility, interrelatedness, fungibility, and interoperability between new and old technologies becomes a major technological constraint;

(8) because the scope and direction of firms' research activity, as well as the incentive to innovate, depend upon the specific features of the tangible and intangible capital stock and the relative weight of irreversible elements it seems clear that the larger the variety of firms with respect to the vintages of their capital structures and the wider the technological spectrum, the more an economic system is able to benefit in terms of the potential for the introduction of radical innovations and the capability of taking advantage of new scientific breakthroughs and related technological opportunities;

(9) the dynamics of irreversibility and innovation can help us understand the emergence of technological clusters, defined as a set of interdependent and complementary technologies which have been gradually introduced along technical paths defined in terms of compatibility between different vintages of irreversible production factors;

(10) market competition also becomes technological competition when it is assumed that different firms may have different endowments of superfixed production factors; the entry of new firms plays a major role in this context in that it provides the opportunity to operate along a technological axis and introduce technological innovations which are not necessarily localized in the limited technical space defined by the incumbents' endowment of superfixed production factors.

The most important conclusion of our analysis is that complementarity among technologies is endogenous to the system.

Complementarity is the result of the localized search for new technologies by firms constrained by irreversible production factors. Complementarity, in fact, is the result of the ability to discover new uses for old inventories, stimulated by changes in the economic environment to which firms cannot react by means of sheer adaptation.

At the end of this analysis, it is clear that the relationship between irreversibility and innovation appears quite complex. According to Salter (1966), in fact, irreversibility delays instantaneous adoption of superior capital goods embodying new technologies and actually explains their diffusion, that is, the time distribution of adoption according to the vintage structure of existing capital stocks. According to this analysis, however, irreversibility stimulates the generation of new localized knowledge and the introduction of new localized technological innovations. The introduction of localized technological change becomes a means and a tool to increase the flexibility of firms coping with fluctuations in both product and factor markets which cannot be accommodated by the existing stocks of installed fixed capital goods and other intangible production factors.

4. IRREVERSIBILITY, TECHNOLOGICAL CHOICES, AND ENDOGENOUS COMPLEMENTARITIES

The irreversibility of the stocks of tangible and intangible capital is not only an important inducement mechanism to generate new technological knowledge and introduce new technologies in a localized technical space. It is also an important sorting device which makes possible technological choices that assess the actual profitability of the introduction of new technology. New technologies can be ranked not only according to their total factor productivity levels, but also with respect to their compatibility and complementarity with superfixed production factors.

Historic constraints are easily converted into incentives. Hence, irreversibility becomes an important focusing device directing the generation of new technological knowledge and the introduction of new technologies along axes of diachronic technological complementarities. Innovation strategies will be directed towards the introduction of new process technologies which make it possible to reuse portions of the existing stock of irreversible capital goods.

Similarly, the irreversibility of tangible and intangible capital stocks can be thought to be at the root of endogenous complementarities. In turn, endogenous complementarities can be both internal to firms and external. In the latter case, they give way to technological externalities, in the former to economies of scope. The direction of research activities for product innovations will be influenced by the incentive to introduce new products that have high levels of complementarity with existing ones in that they can share and reuse both tangible capital goods and intangible assets such as technological competence and localized technological knowledge as well as reputation, brands, and market identification with consumers.

An important distinction can be introduced at this point between internal and external irreversible production factors. Irreversible production factors can be internal to the boundaries of the firm in that they are portions of the stock of tangible and intangible capital. Firms, however, can have access to external and irreversible production factors as well. In this latter case, the inducement mechanism and the related focusing devices, stemming from irreversibility, do take place as well. Firms are now induced to search and innovate, introducing technological innovations with high levels of technological complementarity with respect to such external irreversible factors.

Let us assume that a firm exists and its capitals stocks, both internal and external, include some superfixed production factors. Such superfixed production factors may be reused at low or even no cost. Intangible capital consisting of technological knowledge and reputation in the market-place provides a clear example. We now assume that two new technologies are being introduced in the market-place and become available. The two technologies use the same amount of inputs. The first technology, A, has a higher total factor productivity level but can use only a small portion of the existing stock of 'old' irreversible production factors available in firm i. The second technology, B, has a lower total factor productivity but can take advantage and make new use of a large portion of the existing and irreversible stock of old production factors. From an absolute viewpoint, technology A is superior to technology B, but for the incumbent characterized by irreversible production factors technology B is clearly more productive. More specifically, for some input costs and on the assumption that the portion of irreversible inputs

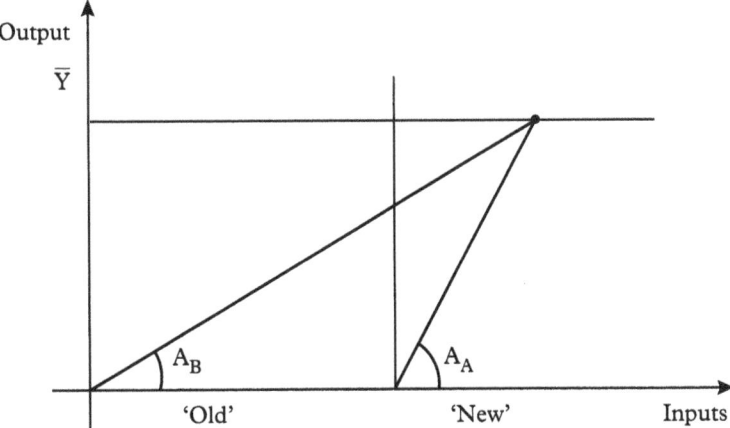

Figure 2.2. *Technological choice: 'old' and 'new' inputs.*

which can be used is free, the technology B is more profitable than technology A. Figure 2.2 provides a simple geometric exposition.

From Figure 2.2 it is clear that all technologies which have a larger total factor productivity level but a smaller coefficient of reuse of existing irreversible production factors are actually inferior for existing firms, but not for new ones. More generally, the larger the amount of superfixed production factors available, the larger is the bias in the technological choice towards new complementary technologies.

The main result of this analysis can be expressed by the traditional negative relationship between cost reduction and the amount of innovation efforts. Localized efforts, defined in terms of the complementarity of new technologies to existing ones, however, in our approach command higher expected results in terms of cost reductions than generic innovation efforts. For a given amount of innovation efforts, firms able to implement a local search for new technologies, a search which is able to make creative use of existing irreversible assets, can obtain larger cost reductions than firms involved in a generic search for new technologies.

This analysis leads to the notion of technological variety. Technological variety here is not the result of the synchronic complementarity among technologies, but rather it is the consequence of

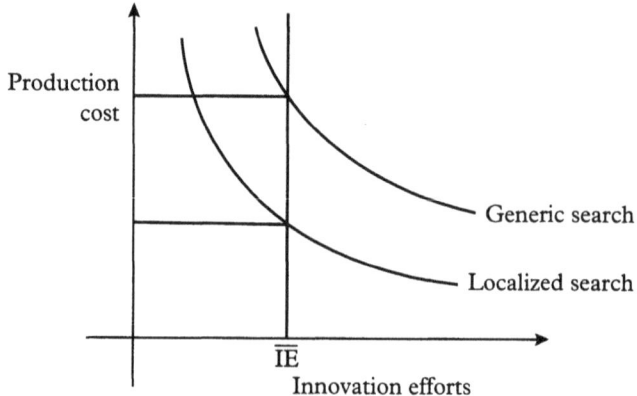

Figure 2.3. *Localized search and production costs.*

the diachronic complementarity (Kirman, 1992). Old technologies exert some local externalities on new ones, provided the new ones can reuse the portions of superfixed production factors already in place.

At each point in time, a variety of different technologies characterized by different levels of total factor productivity and different levels of new use of existing stocks of irreversible production factors can coexist in the market-place according to a given variety of irreversibility of the stocks of existing tangible and intangible production factors in each firm. The intensity of use of irreversible capital stocks compensates for lower total factor productivity and vice versa. According to the endowment of irreversible production factors, firms can select different technologies which can now be ordered according to two factors instead of just total factor productivity levels. The technologies that cannot compensate for lower total factor productivity with higher levels of reuse of existing stocks will not be adopted. By the same token, however, technologies with high total factor productivity but with low levels of compatibility, interoperability, and complementarity with existing stocks of irreversible production factors are not likely to be introduced. As such, they will not be generated and implemented by profit-maximizing firms which are characterized by irreversible production factors.

The constraint and selection mechanism which makes it possible to rank technologies becomes a focusing device. Each firm will try and generate localized technologies that are 'appropriate' to the specific irreversibility conditions of their production factors. The research strategy of each firm will be directed towards the introduction of new technologies that make it possible to increase total factor productivity and reuse large portions of irreversible production factors.

In terms of process innovation, this leads to a clear technological path where firms have a strong incentive to search for technologies that are diachronically coherent with the existing capital stocks so as to implement all latent technological complementarities between old and new technologies, between old capital stocks and new ones. The role of endogenous technological complementarities and interoperability paves the way to understanding economies of scope and externalities as the result of intentional research strategies induced and shaped by the constraints of local irreversibilities.

Economies of scope and externalities stemming from weak indivisibility and technological complementarities can no longer be regarded as an exogenous windfall which benefits some firms able to diversify and or to locate in some districts and operate in some clusters. On the contrary, economies of scope and externalities should now be regarded as the result of a sequence of technological product (and process) innovations which are aligned along high levels of complementarity and interoperability between products and processes introduced by firms characterized by high levels of endowment of irreversible production factors, which are both internal and external to firms.

The introduction of localized innovations induced by irreversibilities leads to a chain of technological complementarities. Such technological complementarities are both synchronic and diachronic, product and process, as well as internal and external to firms.

Synchronic complementarities are defined by weak indivisibilities, and strong compatibilities and interoperabilities among new technologies being introduced. Diachronic complementarities arise when different vintages of capital goods and products are affected by low levels of divisibility and high levels of compatiblity and interoperability. Product complementarities are defined by the joint use of products. Process complementarities by the joint production of specific products. Internal complementarities take place when indivisibilities are found across products and processes that are placed within the borders of a firm. Internal complementarities lead to economies of scope which in turn can be valorized when firms are able either to diversify in lateral markets or to integrate in vertical ones. External complementarities take place among firms and lead to a variety of externalities.

A technological system, as articulated in both technological clusters and technological districts, is defined by the strength of such chains of technological complementarities. In this context, knowledge complementarities play a major role.

5. IRREVERSIBILITY: OUT-OF-EQUILIBRIUM AND ENDOGENOUS GROWTH

The 'now-forgotten models of induced technological change' (Arrow, 2000) showed, long ago, how the introduction of technological innovations can be treated as the result of specific and intentional innovative efforts (Ahmad, 1966; Atkinson and Stiglitz, 1969). In our approach, the inducement mechanism which eventually leads to the actual introduction of innovations is found in the features of the production process where irreversibility matters. However, the market context in which the inducement process is at work plays a strong role, as a selection mechanism and a microeconomic inducement factor.

Any emerging discrepancy between expected and actual factor and product market conditions, associated with investments in superfixed production factors, can be accommodated by established firms with the endogenous generation of localized knowledge and the eventual introduction of localized technological changes along the superfixed production factors endowment axis. The rate and direction of technological change appear to be strongly induced by the share of superfixed production factors in total costs and by the entropy of the business environment. It seems, in fact, clear that the larger are both the former and the latter, the stronger is the inducement to introduce technological changes that change the output elasticity of superfixed production factors and increase general efficiency (Krugman, 1991b).

Our framework of analysis has major macroeconomic implications on three counts. First, our model highlights the dynamic effects of the relationship between expectation, decision-making, and the rate of technological change. Secondly, it has implications

Irreversibility, Innovation 37

for bifurcation in the induced direction of technological change. In fact, when output levels, either for product or factor market fluctuations, change with respect to expectations and are greater than planned, technological changes with a strong fixed-input-saving bias are induced. On the contrary, when output levels, again for either factors or product market changes, differ with respect to expectations but are smaller than planned, technological changes with a strong flexible-input-saving bias are induced.

In the former case, total factor productivity, output, and employment are likely to grow with recursive effects: the increase in employment and productivity engendered by the fixed-input-saving technological change, itself induced by the discrepancy between expectations and actual output levels, is in fact likely to feed further into an unexpected growth of output with a continual dynamic effect. In the former case, instead, productivity growth takes place in a recessive context where employment declines and consequently demand shrinks: hence output levels decline below expected ones and further flexible-input-saving technological change would be induced. Such dynamics of expectations, irreversibility, and biased technological change are thus likely to take place with strong self-reinforcing effects which can account for cyclical growth and provide some economic explanations for the Schumpeterian gales of innovation.

Because the rate and direction of technological change are much influenced by the dynamics of expectations, irreversibility, and localized technological change, the scope and characterization of each agent and each system seems very much shaped by the steps that have been made up until any point in time: hence an intrinsic and long-lasting definition of the technological specialization of countries and firms emerges as the results of these dynamics.

The third macroeconomic issue concerns inflation and interest rates. It seems clear that high levels of rates of interest and inflation, together with high levels of discrepancy between the market price for new capital goods and used ones, lead to higher irreversibility costs and hence a stronger inducement for the introduction of localized technological change. Inflation, however, can be fully endogenous when and if it is the result of the limitations imposed by superfixed production factors to cope with increasing demand levels. Superfixed production factors explain the rise of inflation and hence interest rates which in turn increase the pressure upon firms to cope with changing product and factor market conditions, via the introduction of localized technological change.

At the microeconomic level of analysis, two results seem important. First, technological complementarities and hence technological externalities and economies of scope can no longer be regarded as exogenous. From the product viewpoint, technological complementarities and related technological externalities and economies of scope can now be regarded as the endogenous outcome of intentional research strategies of firms characterized by high levels of intensity of irreversible production factors, which are both internal and external to each firm, and eager to take advantage of their long economic life. Hence, technological complementarities and the related opportunities in terms of technological externalities and economies of scope can be considered to be the outcome of a sequential, intentional, and diachronic complementarity among products and processes. Such complementarity is intentional, both because each firm is induced by irreversibility to generate new technological knowledge and to introduce new technologies that are complementary and compatible with superfixed production factors, and because the choice of a new technology, out of a basket of new technologies supplied in the market-place by third parties, is biased by each firm's endowment of superfixed production factors.

While technological externalities and economies of scope are often regarded as the result of a synchronic (and exogenous) complementarity among products and production processes, we stress the role of diachronic (and endogenous) complementarities. Secondly and most important, the choice of new technologies, both in terms of adoption of exogenous ones and introduction and generation of endogenous ones, is severely biased by the distribution of superfixed production factors in the system and among firms. At each point in time, the technologies being used are not necessarily the best ones in absolute terms, but rather those which are most convenient with respect to the specific and idiosyncratic production and market conditions of incumbents.

Finally and most important, it can now be claimed that irreversibility itself is largely endogenous to an economic system, for it is the result of the discrepancy between expectations and localized technological change. An economic system with low initial levels of irreversibility, but exposed to frequent and repeated 'surprises', is induced to generate fast technological changes in a well-defined technical space that consists in the introduction of new technologies strictly defined in terms of compatibility, interrelatedness, and interoperability with the existing stock of fixed capital. Hence, the larger the discrepancy between expected and actual output and factor costs, the larger will be the complexity and irreversibility of the technical systems in place in each firm and possibly in each economic system.

The actual rate of technological change, for given levels of super-fixed production factors and entropy in the product and factor markets, will be clearly determined by the effective capability of firms to generate new technological knowledge and hence technological innovations. With high innovation costs, incumbents will be unable to face the quasi-losses associated with the introduction of innovation and will eventually decline. With medium innovation costs, incumbents are induced to make only incremental innovations so as to reduce the distance on the endowment axis between A and the intersection between the isocost and endowment lines. In these circumstances, incumbents can only reduce quasi-losses and come closer to best practice. With low innovation costs and large technological opportunities, incumbents can confront emerging quasi-losses with the successful introduction of radical innovations which make possible a quantum leap on the endowment axis and actually go beyond the 'equilibrium' point with a substantial increase of total factor productivity.

Both at the firm and the aggregate level, the conditions for a self-sustained process of technological change and economic growth are now set. The introduction of minor localized technological changes has direct effects on factor markets affecting the relative prices for production inputs. This, in turn, reproduces the conditions for further Farrell inefficiencies, with given irreversibility levels, and hence the inducement mechanism to search for new technologies. These dynamics take on the character of an endogenous and out-of-equilibrium process of technological change. Such a process is all the stronger when the localized and induced technological change leads to substantial total factor productivity growth and makes it possible for each firm to go beyond the point X which simply restores equilibrium conditions.

This is the case when, both at the firm and aggregate levels, location within technological districts and technological clusters plays a key role. Location within a technological system, both in the regional and technological spaces, can increase access to technological opportunities rooted in a specific system of embedded relations and helps to increase the general efficiency of the technological production function 'g' and hence can favour the introduction of actual productivity-enhancing technological innovations. Total factor productivity, in fact, leads to an increase of the demand for the products. This, in turn, induces firms to generate new technological knowledge and further increases in total factor productivity levels. Provided appropriate conditions for the generation of technological knowledge and the introduction of new technologies are available, a sustained process of interacting economic entropy and technological change can continue in time.

6. CONCLUSIONS

Short-term production theory plays an important role in economic analysis: cost theory, the theory of the firm, and the theory of markets are all based upon a short-term production theory. Yet such aspects of production theory are little explored and the relations between the context in which short-term decision-making applies and the determinants of the rate and direction of technological change have received very little attention in economic analysis.

While in equilibrium, short-term and long-term conditions coincide, in out-of-equilibrium conditions firms produce either too much or too little with respect to the expected factor and/or product market conditions and substantial quasi-losses emerge. Such losses are all the more persistent, the less adaptable are fixed production factors and the wider the divergence between expected and actual market conditions. In such a context, firms bear the costs of lack of flexibility. The introduction of new technologies, compatible with the existing capital stock, can become a rational choice, provided technology is considered as endogenous and flexible, that is, it is considered to be the outcome of a local search and generation process.

Such aspects become all the more relevant when superfixed production factors are a pervasive condition in economic systems. Delayed obsolescence, technical interrelatedness, system bottlenecks, long-term obsolescence, limited span of application of intangible assets, location in well-defined regions are all factors that reduce the capability of firms to adjust, even in the long term, significant chunks of their stock of intangible and tangible capital.

In these conditions, all changes in relative prices and desired output levels are likely to induce a significant decline in both output and price efficiency: firms are unable to produce with the 'correct' combination of flexible and fixed production factors. Emergent technical inefficiency becomes all the more cogent with competitive markets with high levels of technical variety: firms with larger shares, of total costs, of superfixed production factors are soon exposed not only to a decrease of technical efficiency but also to the actual decline of market shares and profits.

Technological change interacts significantly with irreversibility both when it is exogenous and hence when diffusion matters and when its endogenous character is recognized. In this latter case, firms can change technologies in order to cope with the constraints of irreversible production factors.

Economic systems with high levels of superfixed production factors are likely to experience higher rates of rigidity and quasi-losses, but also higher rates of introduction of localized innovations, compatible with their endowments of fixed capital. The same countries, however, are likely to experience lower levels of diffusion of technological innovations that are not easily adaptable to existing capital stocks.

Such a close relationship between irreversibility and the inducement to introduce new superior technologies in a specific and localized technical context, as modelled in this chapter, should provide a clear link between path dependence and increasing returns, along the lines insightfully articulated by Arrow (2000). The rest of the analysis in the book is, in fact, now devoted to assessing the specific conditions which make it possible for such dynamic increasing returns actually to take place.

The localized introduction of new technologies induced by irreversibilities lead to the emergence and implementation of technological systems. A technological system can be defined as a set of technological complementarities that are the result of the localized search for new technologies compatible with existing irreversible production factors.

Such technological complementarities can be aligned along three relevant axes. Technological complementarities can be synchronic and diachronic. Complementarities are synchronic when they characterize products and processes that are being introduced at the same time. They are diachronic when they take place among different vintages of products and processes. Technological complementarities can concern products when two or more goods are used jointly; and processes when two or more products are produced jointly.

Finally, technological complementarities can be internal and external to each firm. In the former case, the firm has been able to internalize the technological complementarities and to take advantage in terms of economies of scope. In the latter case, technological complementarities are external and lead to the well-known externalities. Knowledge complementarities in this context are most important as they are both synchronic and diachronic, internal and external, and concern both products and processes.

When and if localized technological knowledge becomes collective within technological systems, such as technological districts and technological clusters, relevant advantages in terms of increasing returns become available.

In such conditions, it is clear that the larger the irreversibility costs, the larger is the likelihood that firms will react by introducing important innovations that make possible significant increases in total factor productivity. We can now turn our attention to the analysis of the conditions for the accumulation of knowledge and the introduction of technological change, with special attention to the role of technological systems in regional and technological space.

In this context, the new theorizing about the collective character of technological knowledge, the incentive system for learning, and the access conditions to local pools of external knowledge and the levels of technological opportunities that stem from the interaction and technological communication among firms and between firms and local research institutions within technological districts and technological clusters play a major role. Analysis of the interactions between irreversibility and technological change provides the basic insight necessary to grasp the elements of a self-sustaining process of endogenous growth fed by the creative reaction of firms whose adjustment to factor and product markets fluctuations is made difficult by superfixed production factors. The quality and characteristics of the technological environment in which firms operate and the conditions for technological communication among firms and other learning organizations, together with the industrial

Irreversibility, Innovation

dynamics, are likely to be the factors most conducive to sustaining fast rates of introduction of new technologies.

The notion of path dependence emerges here as the result of the blending of the localized technological change approach and the broader issue of path dependence. In path dependence, economic action in each step is stochastically influenced by the past, but not deterministically caused by previous events. In a technological path, the probability of introduction of each new technology is contingent upon previous innovations as well as cumulated technological competence, but also on other necessary conditions and constraints such as the levels of irreversibility of the capital stock and the conditions of access to local knowledge externalities. Specifically, local externalities, as opposed to global ones, play a key role in this context. They are made available by location in technological districts and technological clusters that are in a conducive regional, technological, and industrial environment, characterized by qualified interactions with other complementary innovators both upstream and downstream. In this respect, technological path dependence seems able to accommodate within a single framework the causes of both success and failure, so as to explain why some firms innovate more than others, why some firms fail to innovate and actually decline, and to explain the variety of possible innovative outcomes at the system level. The fragile mix of complementary and yet necessary conditions affecting the transition from each step to the next along the path becomes key to understanding the actual sequence of events in assessing the rate and direction of technological change.

3

The Economics of Localized Technological Knowledge: Learning Recombination and Increasing Returns

This chapter briefly introduces the notion of localized technological knowledge and reconsiders its evolution from the Arrovian notion of technological knowledge as public information. It lays down the basic elements and introduces the hypothesis of increasing returns in the production of knowledge. It puts forward the basic conditions for increasing returns to take place: internal learning and access to external knowledge. In so doing, it introduces the analysis of Chapter 4 which is devoted to the role of dynamic efficiency wages and Chapter 5 where the conditions under which localized technological knowledge can become a collective activity are analysed.

1. KNOWLEDGE AS AN OUTPUT AND AN INPUT

The standard microeconomics of innovation has been built upon and implemented on the basis of a number of important assumptions. It has been often assumed that technological information can be treated as an economic good, although one with a strong public character, and that knowledge and technological information coincide. In this context, the generation of technological knowledge has been viewed as the result of a deductive top–down chain that starts from scientific discoveries mainly developed in pure academic research. Eventually such knowledge is applied to the specific activities of each firm (Arrow, 1962b).

Learning is automatically associated with new vintages of fixed capital and intermediary goods, human capital, but rarely with the intentional effort of employees. As such, learning does not require any specific effort and intentional conduct by employees in order to take place (Arrow, 1962*a*). Employees can use new vintages of capital goods embodying technical changes without specific dedicated incentives and employers can freely benefit from such a windfall without any dedicated action or any costs.

Technological information can be appropriated only to a limited extent and spills into the atmosphere: other firms can freely take advantage of it with clear incentives for free-riding and opportunistic behaviour. Intellectual property rights and public subsidies partly reduce the misallocation of resources to inventive activities but cannot overcome the basic inappropriability problems which stem from the public-good character of technological knowledge (Arrow, 1969).

Recent developments in the economics of innovation have questioned this approach on many counts. The view of technological knowledge, as information, is being increasingly challenged and a distinction between information and knowledge is increasingly appreciated.

Technological knowledge is a process activity.[1] It can be exchanged and partly traded in the market-place, as a communication-intensive service, but within a well-specified institutional context. Specifically, knowledge can be exchanged and traded only when interactive communication takes place among vendors and customers. Moreover, in this approach, knowledge cannot be regarded only as an output, but also and mainly as an input: knowledge is mainly an intermediary input which is necessary for the production and consumption of goods as well as for the production of new knowledge.

The notion of localized technological knowledge stresses this dynamic understanding of knowledge as a process activity. In this approach, technological knowledge is considered as an ongoing bottom-up process of problem-solving activities, characterized by substantial indivisibility within modules defined in terms of strong complementarity and hence strong internal and external cumulativeness. In such a context, communication among learning agents

---

[1] It may be interesting to note that in the Latin languages the translations for knowledge (either *sapere* or *conoscenza* in Italian; *savoir* or *connaissance* in French) are respectively a verb or etymologically a derivative of a present participle: in both cases an activity rather than a state.

is an essential condition both for the generation and exchange of technological knowledge.

New appreciative theorizing also based upon new evidence has suggested the demise of the linear model which related unidirectionally scientific progress to technological advances and the decline of the notion of technological knowledge as a bookshelf of blueprints easily available to everybody (Kline and Rosenberg, 1986). In this new approach, technological knowledge is distinct and yet bijective, that is, bidirectionally interactive with scientific knowledge (Metcalfe, 1995a).

Secondly and most important, technological knowledge is now viewed as an indivisible and yet fragmented and dispersed stock of structured information. Because of its highly idiosyncratic applications and specific contexts of implementation, technological knowledge is embedded in a great variety of specific productive and market conditions and partially owned by a wide variety of agents, each of which is able to command a limited portion of it.

Thirdly, following and elaborating upon Simon's contribution (Simon, 1985; Loasby, 1998), innovative capabilities and broadly, the ability to generate new technological knowledge are now considered to rest upon a specific learning capability which draws from diverse knowledge bases and is able to activate a systemic recombination process.

According to this literature, technological knowledge is 'localized' in tacit learning processes that are embedded in the internal background and experience of each innovator and hence highly idiosyncratic and, in product and factor markets conditions, technological knowledge is rooted in regional space, external to each agent but internal to the district and the system of related techniques.

The generation of new knowledge is mainly the outcome of the intentional efforts of innovators drawing on learning processes, which are localized and specific to their individual history and experience and to the spatial and technological context of action. Drawing on the daily routines of learning employees and from the tacit experience of using capital goods, producing and manufacturing, and interacting with both suppliers, customers, and other manufacturers, localized technological knowledge is eventually implemented via the formal activities of R & D (Malerba, 1992).

The capability to innovate appears to be strongly conditioned by both access to external technological information and the

Localized Technological Knowledge 47

technological and scientific knowledge embedded in the environment of the firm, and by the accumulation of tacit knowledge internal to each firm. This new analysis of the innovation process highlights the important distinctions between information, knowledge, and competence and stresses the role of the competence of the firm to manage and valorize the accumulation of technological knowledge building upon internal learning processes and the characteristics of the innovation system in which each firm is embedded.

In turn, innovation systems are both regional and technological: the former consists of the firms co-localized within a region; the latter of the firms active in complementary technologies and products. Technological systems have both a regional and technological dimension which often coincide: firms are both co-localized and active in complementary technologies.

Technological codified information, tacit experience acquired intentionally by means of learning by doing and learning by using in repeated actions by reflective practitioners, and the technological externalities spilling into the regional and technological innovation system, acquired with systematic communication efforts, represent the basic ingredients in the process of the creation of new knowledge. New technological knowledge is a complex process of creation and use of new information, guided by the competence of each agent, building upon the mix of tacit expertise acquired by means of learning processes, the socialization of experience, the recombination of available information, and the conduct of formal R & D activities.

Specifically, according to previous work (Antonelli, 1999a), localized technological knowledge is seen as the result of four distinct processes: internal and external tacit knowledge and internal and external codified knowledge. Internal tacit knowledge consists in skills and rules which 'cannot be articulated'. It is generated by means of processes of learning by doing and learning by using. External tacit knowledge is acquired through informal exchanges and socialization which require dedicated efforts and enable the internalization of the technological externalities spilling into the local innovation systems in which each firm operates. Internal codified knowledge is the result of formal activities of R & D. Finally, external codified information, consisting of structured information available in generic forms, is acquired as such, but eventually reorganized by means of the recombination of bits of technological information and applied in different contexts than those originally conceived. Access to external knowledge is implemented through a variety of procedures: technological outsourcing, formal cooperation schemes between firms with their own R & D laboratories or between firms and universities, including standardization processes, co-localization within technological districts, enhanced division of labour with both customers and providers of parts and components.

In such an activity, each component is indispensable and cannot be disposed of. This clearly implies that no form of knowledge, out of the four considered, should fall below a minimum level without putting at risk the whole process. Substitution can take place, but only to a limited extent. Tacit knowledge is necessary to generate new localized knowledge both directly and indirectly, as it is essential to acquire and learn new codified knowledge, because of the high levels of natural excludability of codified knowledge. Rarely can codified knowledge, even when it consists of the results of scientific undertakings, be reduced to a simple set of instructions. Codified knowledge can be acquired directly on the shelf and yet the direct and intimate relationships between researchers plays a central role in its assimilation. As a result, codified knowledge is necessary to the accumulation and elaboration of both localized knowledge and new tacit knowledge. External codified knowledge is necessary as a source of new ideas to feed the recombination process. External tacit knowledge is required to implement internal knowledge, both tacit and codified. Competence consists in the capability of firms to implement such a complex mix of inputs, where each element is complementary and indispensable.

Because of the complementarity among these distinctive forms of technological knowledge necessary to generate new localized technological knowledge, firms with low levels of intentional organization of learning processes and low or ineffective incentives to frame and valorize learning can fail to generate appropriate levels of technological innovation. Conversely, an effective incentive structure conducive to fast rates of learning and prompt conversion into new knowledge, for given levels of research and development activities, and the systematic and intentional access to external complementary knowledge, both tacit and codified, can generate high levels of technological change.

The access conditions to external knowledge play a key role. The localized character of technological knowledge increases its appropriability but reduces its spontaneous circulation in the economic system. Access to existing knowledge, moreover, is harmed by the relevant communication costs. The complementarity to own technological knowledge and to own the research agenda of each new bit of knowledge and each ongoing research and learning process is a matter of discovery and time-consuming activities. Technological knowledge, in fact, is industry-specific, region-specific, and firm-specific; and because of this it is costly to use it elsewhere: respectively, in other industries, other regions, and other firms (Antonelli, 1999a).

The stronger the codified content and the lower the decodification costs, the greater is the possibility for prospective customers to screen the market-place and assess the relevant bits of knowledge which are actually complementary. Clearly when and where each bit of technological knowledge is kept hidden and obscure by the strategic behaviour of owners worried about low replication costs and high imitation opportunities for third parties and where, moreover, the search, assessment, and decodification costs are high because of its large tacit and idiosyncratic content, access to external knowledge becomes extremely costly and is substantially barred, unless specific contextual conditions apply (Hirschleifer, 1971).

The production of technological knowledge is characterized by both horizontal and vertical indivisibility and systematic cumulability, at least within modules defined by strong complementarities among kinds of knowledge. The generation of new technological knowledges is, in fact, affected by modularity articulated in vertical and horizontal indivisibilities. Vertical indivisibility or cumulability matters in that new technological knowledges are generated mainly using previous technological knowledge, that is, standing on the shoulders of giants. The generation of new technological types of knowledge is also characterized by horizontal indivisibility in that each increase, even with a narrow scope, can have relevant effects in terms of complementarity and additionality with other parallel and yet convergent advances made in seemingly unrelated fields and contexts which, however, belong to the same module (Stephan, 1996).

Because the generation of technological knowledge is characterized by substantial indivisibility, increasing returns can take place.

Major economies of density and scope shape the cost function of innovation activities. In fact, the accumulated stock of competence and technological knowledge acquired by each agent exerts strong intertemporal effects so that average costs decline with the repeated use of such superfixed production factors. By the same token, many innovation costs, that is, the costs of generating additional technological knowledge and extracting, out of it, relevant technological innovations, can be portrayed as incremental costs which add on to existing long-term fixed costs.

Because of limited appropriability, horizontal indivisibility, and systemic cumulability, complementarities matter in the production of technological knowledge.

Following Simon's seminal contributions, one important aspect of the new understanding of the economics of knowledge concerns the modularity or near-decomposability of knowledge. Knowledge can be considered a complex artificial system:

Complex systems can be approached successfully as nearly decomposable systems... in a nearly decomposable system the short run behavior of each of the component subsystems is approximately independent of the short run behavior of the other components;... in the long run the behavior of any one of the components depends in only an aggregate way on the behavior of the other components. (Simon, 1962/1969: 209–10)

The near-decomposability of knowledge stresses the distinction between intra-components and inter-component linkages. According to Simon: 'Intracomponent linkages are generally stronger than intercomponent linkages. This fact has the effect of separating the high-frequency dynamics of a hierarchy—from the low-frequency dynamics—involving interaction among components' (Simon, 1962/1969: 217).

The distinction between local and global knowledge complementarity can be directly derived from Simon's notion of near-decomposability. Local and modular indivisibility of technological knowledges as opposed to global indivisibility seems an important new specification in the economics of knowledge. The basic understanding here is that there is a variety of different and yet related kinds of knowledge which exhibits localized chains of specific complementarities. Within modules, high levels of knowledge viscosity can be detected. In turn, modules of locally strong complementarities can be detected in a sea of weak indivisibilities.

Because technological knowledge is at the same time an output and an input and there are relevant local complementarities among the stocks and flows of knowledge of both each innovator and among connected innovators, substantial local increasing returns are at work. Each bit of knowledge, within modules, adds on to the others, both synchronically and diachronically, and takes advantage of the others with a horizontal and vertical cumulative effect. Economies of scale and density are at work both within firms and among firms, but within modules.

Such local increasing returns can be both internal to the firms, especially when different vintages of knowledge are concerned and the notion of cumulability applies, and external, when local complementarities take place among bits of knowledge stored in different firms. External and local increasing returns are especially important in this context.

Specifically, in this context, both the stock and the increases of local external knowledge become relevant. The generation of each new bit of technological knowledge by each agent requires access to the (fragmented) pool of existing knowledge, which is locally complementary. Moreover, the generation of new knowledge by each agent benefits from instantaneous access to new bits of knowledge, locally complementary, generated by each other agent.

Formally, we can write a production function for the generation of new 'localized' technological knowledge (LTK_i) by each firm i where the internal efforts of research and development activities ($R\&D\&L_i$) complemented by intentional intramural learning[2] are the internal production factor. An important factor in the generation of new technological knowledge for firm i, however, is also the research and development and learning efforts ($R\&D\&L_{n-i}$) of the other firms (n−i) with which the firm has established effective communication systems which make possible reciprocal access, shared research efforts, and the implementation of complementary innovation activities.

---

[2] See Ch. 4.

In turn, both the size of the external research and development activities, external learning, and external stock of co-modular knowledge and its efficiency are a function of the number of agents (N) with which each firm is able to activate technological communication. Such a positive relationship, however, exhibits decreasing returns.

Hence we can write the following 'production activity':

$$LTK_i = j(R\&D\&L_i, EK_{n-i}) \qquad (3.1)$$

where

$$EK_{n-i} = m(N); \quad m' > 0, \text{ and } m'' < 0 \qquad (3.2)$$

External increasing returns take place until a maximum number of firms is engaged in complementary innovative activities and a maximum number of communication channels among them is reached. It is important to note that EK_{n-i} is a factor external to each firm. The number of connected firms is not under the control of any specific authority and can be reached only as a consequence of a discovery process which takes place over time and can be assimilated to the traditional entry process of the Marshallian analysis of competition. External knowledge is relevant not only if it exists, but also if and when an appropriate communication environment is in place. Communication implies a bilateral decision process: its implementation requires that both parties are willing to participate. An intrinsically collective character lies at the core of the analysis.

In other words, increasing returns matter only within well-specified institutional contexts and within modules of technological knowledge and are the result of a long-term discovery process which takes time to be implemented and put in place. Moreover, increasing returns last only until a maximum number of communication channels among firms is reached. The identification, specification, and empirical qualification of the conditions which make it possible actually to achieve such external increasing returns in the production of knowledge constitute one of the main objectives of this book. The rest of the analysis, in fact, tries to understand the determinants and effects of such localized increasing returns.

Major positive feedback-increasing returns are at work in the generation of technological knowledge. Specifically, feedback-increasing returns here stem from the cumulative character of technological knowledge and display their effect over time (David, 1994).

Positive feedback-increasing returns become clear when the role of irreversibility in inducing the actual generation of localized technological knowledge and its eventual translation into localized technological change is recalled. It is important to note that the efficiency of equation (3.1) bears a direct and strong effect on the levels of innovation costs, as stylized in equation (2.2). The larger the efficiency in the generation of technological knowledge, the lower are innovation costs. In turn, the lower the innovation costs, the larger is the amount of technological knowledge a firm is induced to generate to face the irreversibility trap.

Because of the important role of the stock of knowledge in the generation of new knowledge, we see that important positive feedbacks are at play, both internal to firms and external. The larger the amount of innovation activities each firm has funded at time t, the larger are: (1) the incremental amount of internal technological knowledge each firm can add to its own specific stock of localized knowledge and (2) the amount of current and past innovation activities by other firms within technological districts and technological clusters with which firm i has good communication systems in place.

Larger internal and external stocks of knowledge and larger flows of incremental external knowledge exert a direct effect on the amount of technological knowledge each firm can generate at time $t + 1$.

Significant feedbacks are also in place on the communication side. The larger the number of connected firms, that is, firms able to share a common language and interconnect communication channels, the lower are the unit communication costs, provided appropriate communication structures are set in place.

This approach to positive feedback makes it possible to use, in the analysis of the generation of technological knowledge within technological districts and technological clusters at large, the so-called 'Matthew effect' originally elaborated in the sociology of science and eventually generalized to the economics of science and technology by Paul David.

According to Paul David (David, 1998), the Matthew effect consists in the path-dependent dynamics of learning to learn. Scientists who were previously able to achieve a scientific reputation and to access research funds have a greater chance of increasing their competence, scientific output, and eventually productivity with a self-reinforcing mechanism. The larger the competence accumulated and the more widespread the reputation, the greater are the opportunities to attract new research funds which in turn feed the positive dynamics.

The Matthew effect at the technological system level applies because of vertical cumulability of localized technological knowledge and horizontal indivisibility in a context characterized by the irreversibility inducement actually to generate new technological knowledge and apply it in the production process.

The production of knowledge is characterized, within circumscribed regional and technological spaces, by increasing returns because knowledge is at the same time an input and output, characterized by localized horizontal and vertical indivisibility, which takes place within modules of complementarity and interconnection. Major economies of density shape innovation activities within each firm and within groups of connected firms able to share a web of communication channels. In fact, the accumulated stock of competence and technological knowledge acquired by each agent and the web of communication channels exert strong intertemporal effects, so that average costs decline with the repeated use of such superfixed production factors, both internally and externally. As such, the generation of technological knowledge, within modules, is characterized by increasing returns.

Firms can take advantage of such potential increasing returns only when and if a number of conditions apply. First, appropriate systems of interaction among innovators who control bits of complementary knowledge must be put in place. Secondly, appropriate levels of internal accumulation of knowledge must be stimulated.

In this context, learning plays a key role and without the intentional efforts of all the agents participating in the production process, at all levels of the hierarchical structure of the firm, it cannot take place. Learning, as with many other activities in economic theory, cannot be considered a free good or a windfall.

The implementation and valorization of learning requires specific organizational procedures and a specific, dedicated incentives structure. Only when such organization is in place can firms actually learn and the positive effects be transformed into the accumulation of effective localized technological knowledge and the eventual introduction of new technological innovations.

4

Dynamic Efficiency Wages, Learning, and Innovation

Efficiency wages have been defined as wages paid in excess of labour productivity levels and a variety of implications have been detected in a static context of analysis, mainly in terms of information asymmetries and labour market failures. Dynamic efficiency wages are but equilibrium wages for learning firms that are able to elaborate proper incentives for employees to actually learn and participate in the effort to generate new localized technological change and hence introduce new technologies. The equilibrium conditions for 'dynamic efficiency wages' can be detailed when technological change is considered endogenous to learning firms and local innovation systems.

Important wage differentials have always been observed in advanced economic systems. Wages differ greatly across regions, industries, and firms. Growing empirical case-study evidence documents the strong association between wages and innovation activity. This association is observed both at firm and regional levels, especially in the new service economy. The clustering of innovative activity in well-defined local innovation systems and local labour markets with high wages, often well beyond average national levels, and low unemployment levels is found in most empirical analyses. In a parallel way, a strong and positive relationship between innovation intensity, as measured by R & D expenditures, patents, and innovation counts, and wages is observed at the firm level. This relationship appears especially strong in the knowledge-intensive business service industries which now account for the major part of both the rate of introduction of technological innovations and the national product and employment in advanced economies (Gallouj and Weinstein, 1997; Sundbo, 1998).

This chapter addresses the relationship between efficiency wages, localized learning, and innovation in a dynamic and microeconomic context of analysis. It is based upon the hypothesis that technological change is endogenous to the intentional conduct of learning agents. The chapter is structured as follows: Section 1 provides a short account of the debate on efficiency wages. Section 2 presents a model of efficiency wages and total factor productivity growth obtained with the introduction of endogenous product and process innovations. In Section 3, the main implications are discussed.

1. DYNAMIC EFFICIENCY WAGES AND LEARNING

Efficiency wages can be defined as wages that employers pay in excess of short-term labour productivity and hence above market-clearing levels. The notion was first introduced into the economic literature by Solow (1979) and Akerlof (1982) and eventually implemented in a variety of ways. The implementation of this notion has mainly been conducted in a static context of analysis but it paves the way for important dynamic developments.

The basic assumption is that wages paid by employers in excess of market-clearing levels help increase labour productivity but not total factor productivity. Labour productivity can be increased by efficiency wages for a variety of reasons which have been explored by important contributions. Little attention has been paid, on the other hand, to the hypothesis that efficiency wages can stimulate employment involvement. This should enhance learning capabilities and hence contribute to absolute levels of total factor productivity.

Efficiency wages have so far been justified on the grounds of different static hypotheses where the rates of introduction of technological changes and hence of the increase of total factor productivity and the advance of best practice were not considered under the control of the firm. Efficiency wages are instead treated as a tool to reduce or minimize organizational failures that result from the well-known limits facing principals in organizations when assessing all relevant information and monitoring effectively the conduct of agents (Arrow, 1974).[1]

---

[1] Elaborating upon the Arrovian approach, the shirking model first elaborated by Shapiro and Stiglitz (1984) assumes that employees have some discretion concerning their performance and wages paid in excess of market-clearing levels can

The analysis carried out by Akerlof (1982) with his pioneering contribution was in fact much closer to a dynamic interpretation of the causes and effects of efficiency wages. Akerlof elaborated the notion of a partial gift-exchange model where efficiency wages would become a tool to improve the average levels of effort workers are willing to contribute in team production processes, where the actual contribution of each worker is difficult to assess and the average behaviour quickly becomes the norm. As a result, team work can actually become more productive even beyond best practice and engender the introduction of improvements in the production process.

The new evidence about high-wage and innovative firms, mainly active in knowledge-intensive business-service industries, urges us to reconsider the role of efficiency wages as incentives to stimulate employees' involvement and encourage the creative efforts of employees. According to this suggestion, a new approach to understanding efficiency wages can be elaborated, building upon a microeconomic and dynamic approach, one which emphasizes the endogeneity of technological change and its intimate relationship with learning processes. The hypothesis that efficiency wages of a dynamic kind can contribute to the endogenous generation of localized technological knowledge and hence feed the introduction of

help reduce shirking and opportunistic behaviour generally so that workers would actually achieve best-practice levels. Efficiency wages do increase the effort workers are willing to contribute in the production process but this has no effect on the state of technology which is given. Calvo and Wellisz (1979) contribute a more managerial understanding of efficiency wages where employers try and minimize the costs of monitoring the opportunistic behaviour of workers, using efficiency wages as an *ex-ante* incentive to behave properly. According to Weiss (1980), efficiency wages can positively affect labour productivity because it makes it possible for firms to attract more able workers. Efficiency wages become a screening mechanism which makes it easier for employers to select the most appropriate competencies of workers in complex tasks. Salop (1979) provides one more explanation for efficiency wages, focusing on their positive effects in terms of reduction of turnover and consequent losses in labour productivity due to retraining costs associated with the exit of competent workers from each specific function in the production process. According to Salop, workers paid above the market-clearing levels will be more reluctant to leave and turnover rates would be reduced. This short review of the main models available in the literature confirms that efficiency wages are mainly viewed as a managerial tool and practice that employers use to reduce the losses engendered by information asymmetries in the relationship with employees with respect to virtual labour productivity levels.

localized technological innovation can now be spelt out: employers are willing to pay wages in excess of short-term labour productivity levels because they are confident that increased remuneration of labour can be conducive to the actual introduction of new and better technologies.

The rates of generation of localized knowledge depend heavily on learning processes and their cumulativeness. This focuses attention on the role of employees in the most advanced industries, in terms of levels of involvement, active participation in the production process, and intentional efforts directed towards the accumulation of organizational and technological capital.[2]

Active participation and the contribution of the emotional and intellectual efforts of a qualified workforce in implementing learning processes makes it possible to accumulate and better valorize tacit knowledge and experience, enabling the proper evaluation of the specific context of action, and enhancing the match between the availability of new codified knowledge and the experience of each firm. The rates of implementation of know-why, know-how, know-where, and know-when rest on the levels of participation of the skilled workforce in both production and decision-making (Adler, 1992 and Adler and Winograd, 1992).

This relationship seems to play a strong role, especially in knowledge-intensive business services where innovation consists often in the generation of customer-tailored, specific, and highly idiosyncratic solutions. Such innovative solutions can be elaborated only by involved and creative employees who are able to combine generic knowledge and hence high levels of human capital with a dedicated effort actually to understand the specific business conditions of customers (Bessant and Rush, 1995).

In this context, dynamic efficiency wages become especially relevant. Firms set efficiency wages at a level that is in excess of short-term labour productivity levels and the opportunity cost of the effort of workers, in order not only to discourage shirking, but

---

[2] The fast diffusion of a variety of incentives schemes such as management by objective, stock options, stock grants in many firms and at lower and lower levels of the hierarchy within companies can be considered a reliable indicator of the growing attention paid to a direct link between the actual capability of employees to participate in the creative effort of the firm and their actual wages (see Prendergast, 1999).

Dynamic Efficiency Wages 59

also and mainly to exert a direct positive effect on the active participation of the workforce in learning processes. Dynamic efficiency wages enhance loyalty and commitment and stimulate practitioners to develop informal relations and better collective work, sharing information, and accelerating the emergence of tacit knowledge (Aoki, 1988).

Moreover, dynamic efficiency wages exert a strong effect on the levels of mobility of labour in labour markets, both with respect to new generations and interfirm mobility. Dynamic efficiency wages make it possible to attract labour with high levels of human capital and tacit knowledge, with a cream-skimming effect on labour markets, enhancing the socialization of tacit knowledge and more generally the transfer of localized knowledge among firms and between training institutions and firms.

Effective internal labour markets which favour the upgrading of competent employees within the firm are an important complementary tool to accelerate the rates of accumulation of experience and tacit knowledge. In fact, such markets keep competent labour within the firm and act as a powerful incentive to stimulate the participation of the workforce in learning processes. From this viewpoint, dynamic efficiency wages reduce turnover and hence the disposal of tacit knowledge which had been acquired and not yet made available to the organization. Dynamic efficiency wages increase learning rates for they help retain skilled manpower and prolong the time available to the firm to valorize the tacit knowledge acquired by each competent employee and blend it with the codified and tacit knowledge of other members in the organization.

In sum, dynamic efficiency wages stimulate the opportunity for the internal mobility of motivated employees able to participate creatively in problem-solving activities, and activate the inductive processes of learning by doing, learning by using, and most important, learning to interact with customers and in procurement, which can significantly feed the accumulation of localized knowledge.

Dynamic efficiency wages are the incentive to encourage both the bottom–up process of accumulation of competence and innovative capability and the top–down process of adaptation of new codified knowledge to the idiosyncratic context of each firm. Dynamic efficiency wages facilitate the processes of 'translation' of tacit knowledge, acquired by means of learning processes within the workforce, into codified knowledge, and vice versa. In so doing, dynamic efficiency wages and internal labour markets have a powerful effect, accelerating the blending of internal and external knowledge and its integration with the organizational knowledge on which the introduction of localized innovation rests.

These trends in the reorganization of the production of knowledge, with the emerging key role of the creative effort of employees, have been much solicited by the diffusion and implementation of new communication technologies. The interaction between the development of communication technologies and the evolution of the organization of the generation of new knowledge is twofold. The introduction of communication technologies changes the process and the organization of the accumulation of new knowledge and stresses the role of a systematic and structured problem-solving attitude on the part of creative employees. The new conditions for the accumulation of technological knowledge and the elaboration of an appropriate institutional and organizational set-up, able to valorize the creative efforts of employees, in turn, affect the pace and direction of the technological convergence upon which the evolution of communication technologies rests (Antonelli *et al.*, 2000).

Forward-looking firms that are able to become learning organizations command an internal organization geared towards the direct involvement of employees in the accumulation of tacit and codified knowledge, blurring the walls between research and development laboratories and actual production processes in order better to accommodate the insights of employees and integrate them into a continual effort to improve both processes and products.

In manufacturing firms, as well as in knowledge-intensive business services, not only workers in the production line are involved in such activities, but also commercial personnel both in the final and intermediary markets, both in the marketing and in the purchasing departments, so as to be better able to acquire external knowledge and use it directly as an intermediary tool in the innovation process. Learning firms have also implemented a structure of incentives able to increase the innovative effort of employees as a way to stimulate their creative commitment: dynamic efficiency wages.

The distinction between the standard effort, familiar to the debate about efficiency wages, that is, the effort that is necessary to reduce shirking and perform near the efficiency frontier and the creative effort, necessary actually to contribute to the innovation process, seems an important one here.

Dynamic Efficiency Wages

Efficiency wages do increase the effort function of employees and consequently information asymmetries are reduced and possible mistakes in the selection of personnel or in their conduct, together with a variety of organizational failures, can be minimized. The average practice can come closer to virtual productivity levels and coincide with best practice.

Dynamic efficiency wages, paid by learning firms able to command the generation of localized technological knowledge, can increase the creative effort of employees and become a direct contribution to the enhancement of the total factor productivity level of the learning firm as it moves towards the frontier of possible production with the endogenous introduction of product and process innovations. Employees, stimulated by dynamic efficiency wages, contribute to the innovation process and participate directly in the problem-solving activity which generates localized technological knowledge (Ichiniowski et al., 1997; Honig-Haftel and Martin, 1993).

2. DYNAMIC EFFICIENCY WAGES AND THE RATES OF GENERATION OF LOCALIZED KNOWLEDGE: A SIMPLE MODEL

In a dynamic and microeconomic approach to understanding efficiency wages, the actual accumulation of localized technological knowledge rests upon the rate of effective participation of the workforce in the learning process. Such participation and commitment, in turn, is stimulated by dynamic efficiency wages. Hence the hypothesis of a direct positive relationship between the levels of efficiency wages and the levels of total factor productivity can be put forward.

To do this, we can rely upon the classical technology production function introduced by Zvi Griliches (1979) where the parameter of the total factor productivity is endogenized as a function of the research efforts of firms.[3] In our approach, total factor productivity

---

[3] Important contributions anticipating this line of enquiry have been made by Phelps (1966) and Nelson and Phelps (1966) who suggest that total factor productivity, as measured by the term A, depends on the level of employment in research and generally on the human capital of employees. The notion of creative effort, however, is missing here and hence its relationship with the necessary incentives.

levels are portrayed as a function of dynamic efficiency wages, that is, wages fixed in excess of short-term labour productivity levels, but finalized to increase actual total factor productivity via the stimulation of intentional learning, the consequent accumulation of tacit and codified knowledge, and the related introduction of product and process technological innovations.

Efficiency wages, that is, wages beyond the short-term equilibrium levels, however, have also negative implications for profit levels in terms of increased production costs.

The basic conditions for a trade-off between the costs of efficiency wages and their benefits can now be assessed on the basis of a profit equation (4.1) which includes in equation (4.2) a technology production function where total factor productivity levels (A) are an endogenous function (eqn 4.3) of the general capability of the learning firm to generate localized technological knowledge (LTK) and hence introduce technological innovations. The positive and strong effects of dynamic efficiency wages upon such technological capability, in turn, are made explicit by equation (4.4).[4] Total costs are also influenced by dynamic efficiency wages (WL) which are now considered as an addendum on the cost side, next to capital costs (RK). Formally the model is specified as follows:

$$P = RT(Y) - CT(Y) \tag{4.1}$$

$$Y = A(K^a L^b) \tag{4.2}$$

$$A = f(LTK) \tag{4.3}$$

$$LTK = W^c \tag{4.4}$$

$$CT = RK + WL \tag{4.5}$$

The equilibrium conditions for W and Y can be determined by setting equal to zero the partial derivatives of the profit function based upon the dual of the production function and the revenue equation which makes explicit the positive role of efficiency wages upon total factor productivity levels. After substitution of equations (4.4) and (4.3) into (4.2) we can write:[5]

$$Y = (W^c)(K^a L^b) \tag{4.6}$$

---

[4] Equation (4.4) can be considered a reduced form of equation (3.1) where it is assumed that internal learning (L_i) is directly stimulated by W^c.

[5] For equilibrium needs we assume $0 < c < 1$.

Dynamic Efficiency Wages

The cost function which embodies the effects of dynamic efficiency wages is the dual of equation (4.5), which, with standard assumptions about constant returns to scale, can easily be written as follows:

$$C = Y/W^c (R/a)^a (W/b)^b \qquad (4.7)$$

The revenue equation for a price-taker firm in a competitive market, hence with given and inelastic price with respect to both output and costs, is simply equation (4.6).

Setting the partial derivatives of the revenue and cost equations with respect to wages equal to zero yields:

$$dY/dW = cW^{c-1} K^a L^b \qquad (4.8)$$

$$dC/dW = Y/cW^{c-1} (R/a)^a b^{1-b} W^{b-1} \qquad (4.9)$$

In equilibrium the optimum level of wages is:

$$W = (cb^b / (R/a)^a (b/c))^{1/a-c} \qquad (4.10)$$

Equation (4.10) confirms that wages will be larger the larger is c, that is, the greater the effect of learning upon total factor productivity levels. Equations (4.9) and (4.10) show that wages are dynamic efficiency wages, that is, wages in excess of the 'static' levels of labour productivity, but still equilibrium wages when their 'dynamic' effects upon total factor productivity are taken into account.

The solution of the standard profit maximization procedure is useful in many ways. It makes it clear that:

(1) the larger the effect of efficiency wages on total factor productivity, the higher is the equilibrium level of wages;
(2) hence and consequently, dynamic efficiency wages are above market-clearing levels only in the short term and not in absolute terms;
(3) specifically, dynamic efficiency wages are above market-clearing levels only when and if firms are not able to valorize the intentional and dedicated efforts of workers in the production process as an intentional and systematic part of a more comprehensive innovation process;
(4) the greater and the better is access to external knowledge and the better are conditions for technological communication

within local innovation systems and technological clusters, the greater the general effects of efficiency wages on current total factor productivity levels and hence on the actual equilibrium levels of wages, when taking into account the effects on total factor productivity levels;

(5) if the relationship between dynamic efficiency wages and total factor productivity levels holds, the schedule of long-term labour demand is affected. It will be much more elastic than currently assumed: all increases in wages, paid by learning firms, help shift the short-term labour demand curve to the right. This rightward displacement takes place not only because of the traditional 'long-term' substitution of capital for labour, but also because of the increase in total factor productivity levels and on top of this because of the increase in output due to increased cost efficiency.

These results shed some light also on the debate on the macroeconomic effects of efficiency wages. When technological knowledge is viewed as the endogenous outcome of a bottom–up process of accumulation of competence in which learning plays a key role and appropriate incentives are necessary in order for learning to take place and be implemented through the actual introduction of technological innovations, higher efficiency wages translate into higher total factor productivity which should lead to lower prices, higher output, greater equilibrium demand, and hence higher employment.

Efficiency wages are likely to increase employment rather than reducing it, provided they are 'dynamic efficiency wages', that is, they are effectively paid by 'learning firms' able to capitalize on learning and use it as an intentional input geared towards the accumulation of new technological knowledge and eventual technological innovations.

The levels of wages of an economic system are strongly influenced by the diffusion of managerial techniques which are able to build upon the distinctive innovative competence so as to stimulate the involvement of employees and integrate actively their creative effort into innovative strategies, via efficiency wages, as a tool to accumulate localized technological knowledge and accelerate the pace of introduction of technological innovations.

Because learning firms' conditions of access to external knowledge play a major role in the actual amount of technological knowledge a learning firm can extract from intentional learning, the levels of efficiency wages an economic system can afford are larger, the more conducive is the local innovation system to valorizing internal learning efforts and the better are the competencies of firms at implementing learning activities and finalizing them to introduce technological innovations.

The conditions which favour technological communication within innovation systems play an important role in this context. Technological externalities, spilling from universities to firms and between firms themselves, as a result of accelerated communication flows of both tacit and codified knowledge in innovation systems within well-defined geographical and technological domains, provide the basic engine for the emergence of important positive feedbacks, external to each firm but internal to the system. Organization at the firm in terms of the architecture and effectiveness of communication flows between units within each firm and between firms, universities, and other research institutions within technological and regional innovations systems, becomes a central factor in taking advantage of the scope for the accumulation of knowledge and its timely conversion into technological innovations (Antonelli, 1999a).

3. CONCLUSIONS

Significant wage differentials have always been observed across regions, industries, and firms. Growing empirical case-study evidence documents the strong association between wages and innovation activity.

The focus of innovative activity is shifting away from manufacturing industries towards the new knowledge-intensive business-service industries. The new locus of innovative activity presents specific features which highlight the key role of motivated employees' creative efforts in the accumulation of localized technological and organizational knowledge and in the eventual introduction of technological innovations which are more and more sold as disembodied knowledge-intensive products, specifically tailored to the distinctive needs of heterogeneous customers.

In this context, the role of localized technological knowledge gathers momentum. The accumulation of localized tacit knowledge and skills, embedded in the processes of learning by doing, by using, and most of all learning to interact within marketing and procurement, plays a central role in the generation of new technological knowledge and hence in the eventual introduction of technological innovations, with the final outcome of the enhancement of total factor productivity levels both for producers and users.

Learning, however, is not a 'free lunch' or a windfall. For learning to take place and become an effective input into the generation of new technological knowledge and eventual technological change, it has to be intentionally implemented and organized by firms which are able to blend it with external knowledge and formal research and development activities. Well-defined incentives are necessary in order to retain skilled manpower with high levels of on-the-job-training, to stimulate the effective participation of workers in the learning process, to direct their efforts towards the introduction of actual improvements, not only to reduce shirking, but actually to gear the effort function into being a direct input in the accumulation of tacit and eventually codified knowledge. Wages can be paid in excess of apparent labour productivity by firms which are able to command the learning process and to implement it in order to extract new localized technological knowledge and hence introduce technological innovations, specifically tailored to the needs of heterogeneous customers.

Firms able to command the accumulation of localized technological knowledge and to make creative use of the actual involvement of employees can afford dynamic efficiency wages and can introduce technological innovations at a faster pace than competitors. The active integration of workers in the innovative process, conceived as an organized problem-solving activity, becomes a distinctive management competence which is able to provide a rationale for the strong association between wages and innovation activity, often found in the growing empirical case-study literature.

Such competence is seen as the basis for the competitive struggle, especially in knowledge-intensive business-service industries, where at each point in time innovation and product coincide in the introduction of highly customized, tailored, idiosyncratic solutions to the specific technological and organizational problems of heterogeneous customers, mainly achieved through systematic problem-solving activities.

Only motivated employees who bring together high levels of human capital and creative effort can contribute to the problem-solving activity which lies behind the accumulation of localized technological knowledge and the identification of innovative solutions and generate new specific interfaces between their generic knowledge and the idiosyncratic context of application. Dynamic efficiency wages become an essential component of an integrated approach to industrial relations and market strategy.

From this viewpoint, the analysis of the role of dynamic efficiency wages in the conduct of firms, primarily in knowledge-intensive business-services industries, can contribute a deeper understanding of the dynamics of the generation of new technological and organizational knowledge and the related introduction of innovations in services. This, in fact, is an area of growing importance both with respect to the present rate and direction of technological change and to the most dynamic sectors in terms of employment and output in the advanced economies at large.

At a more aggregate level, our analysis seems to provide a rationale for understanding wage differentials across industries: more innovative industries exhibit higher wages than less innovative ones. At the regional level, dynamic efficiency wages shed some insight on the virtuous cycle between wages and innovation within technological districts. Within such local innovation systems in fact, the larger the spillovers and hence opportunities for socialization and recombination with appropriate technological communication conditions and the more effective the generation of localized technological knowledge based upon active learning, the higher can be the rate of introduction of innovations and hence total factor productivity levels, 'efficiency' wages, and hence output and employment.

Local innovation systems, conceived as communication systems, provide important opportunities in terms of augmented flows of effective Marshallian technological externalities for learning firms. Firms located within such technological districts can better implement their internal innovation capabilities and actually increase their rates of accumulation of tacit and codified knowledge and eventually introduce actual technological innovations with positive effect on total factor productivity levels. In this context, dynamic efficiency wages can become effective tools to accelerate the rate of technological change.

The argument can be spelt out with stronger emphasis when the quality of technological communication is presented as a necessary condition for firms to combine dynamic efficiency wages with high levels of generation of localized technological change and hence fast rates of technological innovation and the effective increase of total factor productivity.

When technological communication is ineffective, firms cannot commit themselves to a strategy of accelerated learning, cannot afford dynamic efficiency wages, and are locked into low levels of total factor productivity. The distribution of technological districts in an economic system and the endowment of effective local innovation systems become major conditions for achieving fast rates of growth of output, employment, and productivity.

Growing case-study empirical evidence about the strong association between wage levels and the pace of innovative activity, and specifically the clustering of innovative activity in well-defined local innovation systems characterized by the concentration of knowledge-intensive business-service industries and local labour markets with high efficiency wages and low unemployment levels across Europe, finds a theoretical explanation and understanding in our analysis.

In sum, when the endogeneity of technological knowledge and hence of the pace of technological change is integrated into microeconomic analysis, it is clear that dynamic efficiency wages, if coupled with a conducive local innovation system and the command of managerial competence, can activate fast rates of introduction of technological change which keep labour productivity levels beyond static equilibria for the incentives they provide to stimulate employees' creative efforts.

This relationship seems to play a growing role in the new service economy in assessing the pace and direction of technological innovations in the knowledge-intensive business-services industries, which, in turn, play a key role, as providers of productivity-enhancing products, to the rest of the economic system.

5

Localized Technological Knowledge as a Collective Activity

This chapter explores the conditions under which localized technological knowledge can acquire collective properties. In Section 1, the role of absorption and assimilation efforts to take advantage of external knowledge are explored. The conditions under which collective knowledge can exhibit increasing returns are identified. Section 2 elaborates the distinction between technological knowledge as a collective activity and the public-good tradition of analysis. Section 3 shows how the notion of technological knowledge as a collective activity makes it possible to integrate the Marshallian tradition of analysis, based upon externalities, and the Coasian approach to transaction costs. Section 4 stresses the role of the collective commons of technological knowledge and puts in perspective the role of technological communication in a systemic analysis of technological knowledge and technological change.

1. LOCAL EXTERNALITIES, ABSORPTION COSTS, AND COLLECTIVE KNOWLEDGE

External technological knowledge is an essential input into the production of new knowledge. Yet significant efforts have to be made by each agent eager to access external knowledge and specific conditions must apply for external knowledge to be effectively used as an intermediary input. Relevant search, decodification, and assessment costs may induce each agent to reproduce internally the necessary knowledge and miss basic technological opportunities with a substantial decline in the overall efficiency of the activities leading to the generation of new technological knowledge. The failure to internalize external knowledge is all the more dangerous when it impedes firms from taking advantage of the increasing returns available in the recombination of external knowledge with internal knowledge.

Consideration of the absorption and assimilation costs incurred in the use of external localized technological knowledge makes it possible to qualify the conditions under which technological knowledge can exhibit the features of a collective activity. A collective activity is found when the production process is shaped by radical indivisibility and hence complementarity of inputs so that property rights cannot be (fully) assigned.

In this context, the costs and advantages of searching for, locating, and accessing the relevant external knowledge play a major role. External increasing returns, stemming from access to a dispersed and yet indivisible stock of existing knowledge which belongs to the same module and to the efforts of complementary innovators, seem especially relevant. External increasing returns can take place when and if interactive and mutual access to common pools of technological knowledge takes place. Analysis of the specific and idiosyncratic conditions for external increasing returns in the production of knowledge becomes crucial.

The basic argument here is that in a world where nobody can claim full control of all existing knowledge, each agent possesses diverse and yet complementary pieces of information and knowledge which are not only useful *per se*, that is, in the dedicated activity in the course of which they have been implemented and elaborated, but also for broader and different uses (Richardson, 1998; Albin, 1998). In a parallel way, it is clear that each individual advance is not only useful for the specific dedicated purpose for which it has been elaborated but also for a variety of other possible uses. This latter argument is relevant in many ways: specifically, each piece of information is useful both in a positive way in that it enlarges the amount of knowledge available and in a negative way in that it contributes to the understanding of dead ends and blind avenues which helps to reduce the waste of resources by third parties.

The notion of localized knowledge is relevant in this context. And our analysis applies to the conditions for the generation of new localized knowledge. The access to fully codified and generic, scientific knowledge, in fact, is, to some extent, provided by universities and academic training which perform the role of maintaining the pool of basic scientific and codified knowledge and making it available to students. Universities, moreover, also provide basic access to their research activities through the institution of publications. Access to technological knowledge, on the other hand, especially when its localized character is stressed, is made difficult by its tacit and idiosyncratic character. Yet technological knowledge and technological procedures elaborated by company A to solve a specific and highly idiosyncratic problem in a completely different and apparently unrelated context, can be of great help to company B, active in another industry and even in a different technological field.

In this context, it is clear that the productivity of that piece of knowledge could be greatly enhanced when and if each agent were ready to put in a common pool each piece of knowledge which is, in fact, complementary to a variety of others. More generally, the generation of knowledge can provide the archetypal evidence of a network process where the productivity of the resources is larger, the larger the number of agents that take part in it. In other words, an issue of network externalities on the supply side arises when and if the basic conditions for the productivity of a specific asset are a function of the number of complementary pieces each other agent is ready to contribute to the (collective) undertaking. In these conditions, the production of knowledge is characterized by increasing returns which are contingent upon the conditions of the local innovation system.

2. TECHNOLOGICAL KNOWLEDGE: PUBLIC GOOD OR COLLECTIVE COMMONS?

Along these lines of analysis, growing evidence emerges of the collective character of technological knowledge. Technological knowledge is collective when and if it is the result of a collective undertaking, in that its generation is the result of a process that combines specific pieces of information and knowledge that are owned by a variety of parties and cannot be traded as such. From this viewpoint, identification of the collective character is related to a process activity such as technological knowledge, that is, the production process of a new service, rather than to the allocation of a good.

This approach to the economics of technological knowledge and the new emphasis on the key role of existing knowledge, viewed as an essential intermediary input, make it possible to reconsider the notion of technological information as a public good.

The public character of technological information emerges in the context of allocation analysis. According to a long-standing tradition, technological innovation should be considered a public good in that 'one man's consumption does not reduce some other man's contribution' (Samuelson, 1954, 1955). Following Arrow's seminal contribution, technological information in fact exhibits the classical features of non-excludability and non-rivalry in use and as such is difficult to appropriate. The definition of technological information as a public good is clearly related to its conditions of distribution and usage, rather than to the conditions of the generation process. In the Arrovian tradition, technological information is public because it cannot be appropriated by the vendor. Our problem, by contrast, is how the perspective user can actually access technological information which is dispersed and often tacit. It is partly appropriated by intellectual property rights and yet belongs to an indivisible set.

The tradition of analysis of technological information as a public good is embedded in allocation analysis while our approach to technological knowledge as a collective activity builds upon the appreciation of the variety and yet complementarity of technological knowledges. Secondly, this approach is the result of a shift in the focus of the analysis from the allocation of technological knowledge to the generation process of new pieces of technological knowledge where existing pieces of technological knowledge are seen mainly as inputs into the generation of new technological knowledge. Existing pieces of technological knowledge, however, are part of an indivisible and yet fragmented production factor whose command and ownership is dispersed in the economic system.

This divide is all the more clear within the localized technological knowledge tradition of analysis. In this context, in fact, new technological knowledge is more and more viewed as the result of the combination of different and yet complementary pieces of technological knowledge. Technological knowledge can be partly appropriated not only by means of intellectual property rights, but also because of its embeddedness in the idiosyncratic production and market conditions of innovators. Localized technological knowledge, because of its embeddedness in specific conditions, exhibits high levels of modularity: bits of knowledge are highly complementary within modules, but less so across the full range

Localized Knowledge, Collective Activity 73

of knowledge. Cumulability is also important in that it is once more local rather than global: new bits of knowledge add on to pre-existing ones within modules that are localized both in regional and technical space. Each piece of technological knowledge, moreover, is rivalrous in that imitators do reduce the quasi-rents of innovators.

Localized technological knowledge can be also better exchanged within barter-based contexts and when repeated transactions are possible. Post-sale assistance, moreover, plays a key role: customers can pay on the basis of the actual advantages provided by the application of the new knowledge. Vendors can better control the actual applications and potential developments of their proprietary knowledge. Both barter-based transactions and repeated post-sale-based interactions make it possible to shift from an *ex-ante* perspective to an *ex-post* one, often based upon an ongoing and revolving contractual context. Trade needs to be implemented by means of non-arm's-length relationships. Also from the tradability viewpoint, we see that the generation of localized technological knowledge has all the characteristics of a shared-process activity.

Technological knowledge can become collective when attention is paid to the special conditions that are conducive to localized increasing returns consisting in the positive effects on the efficiency of the research activity of each agent of the complementary current and past activities of others.

The difference with respect to the public-good approach is sharp in that it should be clear that a new piece of knowledge, generated by means of existing, indivisible, and yet dispersed technological knowledge—made accessible by specific contextual conditions—can itself be partly appropriated because of high levels of reproduction and imitation costs. In this case, a new piece of knowledge has been generated through the contribution of a variety of knowledge elements, each of which was owned and controlled by a variety of agents, but once generated it can be partly appropriated by the agent who has been able to master the generation process. Hence we have the situation where a piece of knowledge can be collective and quasi-public, as well as collective and public. In the former case, a piece of technological knowledge is the result of the recombination of a variety of preliminary elements and bits of knowledge dispersed in the economic and technical system and once produced, it cannot be appropriated.

Within this framework, the collective nature of technological knowledge stresses the importance of the conditions for accessing the pieces of technological knowledge already stored, but dispersed, because they are specified in a myriad of applications and idiosyncratic developments. These conditions are key to qualifying the context in which increasing returns in the production of knowledge can actually take place and hence become a key factor in improving the rate of technological advance in any economic system (Carter, 1989; Arora and Gambardella, 1994).

In sum, localized pieces of knowledge are viewed as distributed across highly differentiated baskets or modules of distinct bits of knowledge, each of which, however, provides the basic, indivisible, and single intermediary inputs into the production process of new pieces of knowledge. Hence, knowledge has the typical codified characteristic of public goods and yet it is dispersed and localized in a variety of specific and localized contexts of application and is partly appropriated by a myriad of users. These technological pieces of knowledge can become collective when and if dedicated efforts are made and institutional conditions apply, so as to make real the implicit and potential complementarity of the different bits. Technological knowledge can become collective when and if actual technological communication takes place among learning agents. Access to collective knowledge depends on the extent to which effective communication among innovators takes place through the innovation system.

In this context, the properties of regional economic systems, conceived as rooted communication networks into which information flows, and technological innovation systems, defined in terms of modules of actual complementarity of the technological knowledge possessed by each agent, matter in explaining the capability of each agent to generate new technological knowledge taking advantage of the localized commons of collective knowledge and hence in conditions of increasing returns.

The conditions for technological communication become relevant in this context and their assessment contributes significantly to the analysis of regional innovation systems (Freeman, 1991, 1997; Nelson, 1993; Antonelli, 1999*a*).

Localization within regional innovation systems characterized by effective communication channels plays a major role in this context. Agglomeration favours interaction and repeated exchanges; reduces opportunistic behaviour and free riding. Agglomeration reduces transactions costs associated with the absorption of technological externalities and favours technological communication (Lundvall, 1985; Von Hippel, 1988; Utterback, 1994; Lamberton, 1996; Engelbrecht, 1998; Antonelli, 2000a; 2000b; Jaffe et al., 1993; Feldman, 1994; Patel, 1995; Swann et al., 1998).

Access to external knowledge within footloose technological innovation systems requires the creation of more explicit forms of cooperation among remote firms. Also in this case, however, technological communication is relevant and can take place through a variety of footloose communication mechanisms such as interfirm mobility, subcontracting and procurement, and user–producer interactions. The frequent overlapping of technological and regional innovation systems and the related overlapping of the technological and regional concentration of innovation activities suggests that these two distinct notions of innovation systems coincide to some extent.

3. COLLECTIVE TECHNOLOGICAL KNOWLEDGE, EXTERNALITIES, AND TRANSACTION COSTS

Our approach to localized technological knowledge as a potential collective activity which can exhibit increasing returns, makes it possible to integrate two traditions of analysis which have contributed to this field: the transaction and the externality schools, respectively.

The externality approach stresses the role of increasing returns within circumscribed regional spaces to which firms have access because of the important role of proximity. Externalities stem from imperfect divisibilities among production factors: proximity provides enhanced opportunities for agents to internalize their benefits (Brusco, 1982; Antonelli, 1986a; Becattini, 1987; Camagni, 1991, 1999).

The transaction costs approach, on the other hand, values the role of proximity in terms of the enhanced confidence and trust that make possible the reduction of the costs involved in the definition of a proper price for goods that have already been manufactured (Storper and Harrison, 1991; Harrison, 1992).

Although the two approaches are often mingled in most analyses, it seems important to stress that they refer to radically different analytical frameworks. The externality approach, in fact, has been elaborated to accommodate increasing returns. Increasing returns are at odds with spontaneous equilibrium. On the other hand, the transaction cost approach identifies such local systems as 'perfect markets' where no market failure takes place and *ex-post* coordination is perfectly carried out by markets. An effort seems necessary to provide a single integrated framework which actually combines the two approaches and yet is able to appreciate their distinctive features.

The integration of the externality approach and the transaction cost approaches becomes productive in order to explain why and how some economic environments are more conducive to fast rates of introduction of technological change, for a given amount of dedicated resources, than others. Technological change is, in fact, primarily generated by technological knowledge which in turn is heavily influenced by the conditions of access to the indivisible and yet fragmented pool of existing knowledge for each agent.

Let us first consider the transaction costs approach. The markets for technological knowledge perform very poorly: trade is made difficult by many transaction costs. First, the quasi-public-good character of existing technological knowledge and the related well-known appropriability problems play a major role here. Agents are reluctant to make access to their own bits of knowledge easy because this would further reduce their appropriability conditions, especially for competitors and perspective imitators. Applications of technological knowledge are rivalrous within industries. The stronger the appropriability conditions of the existing technological knowledge owned by each agent, and the larger the possibility of trading it in the market-place without any risk that access is restricted and rationed, the easier the conditions for both vendor and customer to meet in the market-place and fix a price. On the other hand, however, the owners of each bit of knowledge are rarely aware of the value of their own portion of knowledge for other users who are not strict competitors and, as such, prospective imitators.

This analysis urges reconsideration of the traditional trade-off about intellectual property rights with reference to the twin tragedies of the commons. The first tragedy of the commons constitutes a basic tenet of standard analysis. Collective property leads to excessive use of common resources. Overutilization of given resources is the result of failure to assign property rights. The definition of property rights over indivisible resources and excess zoning can, however, generate other inefficiencies in terms of rationing, such as have often been shown by the circumscribing of property rights on watering and the consequent misallocation of water (Stiglitz, 1994). In a strong intellectual property-right regime which does not care for the implicit risks of technology-rationing, the owners of technological knowledge, in the absence of a fully articulated market for technological knowledge, may place a discretionary limit on access to their knowledge and ration all technological sales, with clear costs in terms of duplication and missed output in terms of unexploited increasing returns. On the other hand, a property-right regime which does not provide any protection is likely to induce industrial secrecy with evident costs in terms of communication and search costs for prospective users. An intellectual property-right regime designed to enforce both appropriability and derivative usage seems, in this context of analysis, necessary. Derivative property rights on knowledge generated by means of proprietary knowledge on the one hand, or copyright-oriented intellectual property-right regimes which reduce excludability but enforce the remuneration of owners on the other hand, might become useful solutions. Thurow (1999) has recently aired the growing concern about the need for a change in intellectual property regimes, stressing at the same time the need to better enforce appropriability and yet reduce excludability, suggesting that different classes of patents could be created. Technological knowledge with a strong generic content, with scope for wide applicability to many different economic activities, could be patented but with the obligation of compulsory licensing. Extensions of existing knowledge with a stronger idiosyncratic content could be patented with high levels of excludability (see also Scherer, 1999).

The notion of collective activities is relevant here in that it entails also, and strongly, the case for externalities. Externalities, as is well known, emerge from imperfect divisibilities of production factors. The useful distinction introduced by Griliches (1992), building upon Scitowsky (1954), between rent technological externalities, that is, pecuniary externalities for which external knowledge is actually purchased at low(er) prices with significant consumers' surplus, and knowledge externalities for which technological information is available in the atmosphere, seems very useful. In our communication approach, knowledge technological externalities matter as much as rent technological externalities.

The distinction between local and global knowledge externalities, however, becomes central in this context. Knowledge externalities are global when information spills into the atmosphere with such intensity and strength that space, both regional space and technical space, does not matter. Everybody, independently of his or her location, in both spaces, is able to receive the information and assimilate it. Knowledge externalities are local, in regional and technical space, when decay with distance and proximity is relevant. When localized technological knowledge matters, and its idiosyncratic and specific character is stressed, knowledge externalities tend to be local rather than global and the relevant communication costs have to be assessed. Externalities spill within a limited region.

In a similar way, the notion of localized technological knowledge, and the related rich empirical evidence on contextual knowledge, makes it possible to introduce the distinction between modular indivisibility as opposed to generalized indivisibility. The Arrovian tradition assumes that knowledge indivisibilities are general so that each element of knowledge belongs to a single body. In the localized knowledge tradition of analysis, by contrast, a plurality of knowledge is identified. Following the notion of complex systems articulated in subsystems introduced by Simon, knowledge can be thought as a complex system where knowledge subsystems relate to each other with strong and weak indivisibilities, respectively. Pieces of knowledge characterized by strong indivisibilities belong to the same modules. A chain of weak indivisibilities relates each module or cluster of knowledge to each other. The notion of modular indivisibility is important because it leads to an appreciation of the clusters of technological knowledge and technological innovations which impinge upon a common of knowledge characterized by strong indivisibilities. Consequently, knowledge externalities are very important within modules, but less strong across modules. Hence we can introduce the distinction between modular externalities and generic externalities.

The structure of communication systems in place, within regional and technological systems, becomes the object of strong interest for economic analysis when externalities are local as opposed to global.

Our main point here (see Ch. 3) is that knowledge complementarities and externalities are much more local, as opposed to

Table 5.1. *Knowledge: public, private, or collective*

Properties	Public good	Private good	Collective activity
Cumulability	full	none	local
Appropriability	none	full	partial
Excludability	none	full	partial
Rivalry	none	full	within industries
Tradability	none	full	barter
Divisibility	none	full	within modules
Externality	global	none	local
User costs	low	low	high

global, than currently assumed. Knowledge complementarities are found more within subsystems than among subsystems (Simon, 1962/1969). Hence externalities spill locally, rather than globally. Within subsystems firms can prevent the use of their own technological knowledge by other parties, only to a limited extent. Within modules there is substantial viscosity and this declines with distance.

The ray, into both technological and regional spaces, where such local spillovers take place is, however, limited and beyond it, technological signals cannot be 'heard'. Because each firm whispers its technological knowledge, other firms need to listen very carefully and build appropriate acoustic instruments.

Table 5.1 summarizes the main results of the analysis developed so far.

These arguments can now be summarized around two points. First, it is now possible to introduce the distinction between social, private, and collective knowledge output. According to Arrow (1962a, 1969), firms invest in the production of technological knowledge less than socially required in terms of welfare because they can appropriate only a fraction of its value. The lower the appropriability conditions, the greater is the divarication between private and social welfare output. However, when the role of technological knowledge as an input, as well as an output, is appreciated and knowledge externalities are accounted for, a collective output can be identified. Now the collective knowledge output function is higher than the Arrovian private one. Secondly and most important, the larger the number of firms contributing to the collective and local undertaking, the larger the collective knowledge output.

The notion of local externalities makes it possible to assess the relationship between knowledge externalities and the number of firms engaged in complementary innovation activities. The cumulability and fungibility of local external technological knowledge to the innovation activities of each firm is bound by decreasing returns. Each agent contributes to the creation of a collective cumulative stock of potential knowledge externalities, albeit at a decreasing rate. In this situation, the relationship between external knowledge and the number of firms is shaped by a positive first derivative and a negative second derivative.

The assessment and analysis of the conditions for access to and effective use of the collective stock of external knowledge now becomes the main issue. External knowledge can provide knowledge externalities, within modules, provided prospective users are able to identify, access, and make a consensual creative use of it. So far external knowledge is likely to provide potential knowledge externalities in a well-defined institutional context. Analysis of the communication conditions which make such potential knowledge externalities actual is developed in the following chapter.

4. CONCLUSIONS

The notion of collective knowledge makes it possible to reconsider the tragedies of the commons. When collective knowledge matters, the access of each agent has a twin effect: on the one hand, it implies access to and use of common resources, but on the other hand, it helps to increase the common pool. Localization within technological districts can be considered as an institutional device, in the context of which firms have intentionally accepted a *de-facto* reduction in their control of proprietary technological knowledge provided an implicit context for the exchange, mutual transfer, and possible trade of technological knowledge becomes accessible. In such contexts, barter exchanges of information within systematic flows of technological communication both complement and substitute for trade. The integration of the transaction costs approach and the externality analysis becomes possible: external increasing returns take place when transaction costs are low and appropriate tools for the institutional governance of communication flows have been designed and implemented.

6

The Economics of Technological Communication in Technological Districts

The institutional context of regional economic systems in terms of communication conditions plays a major part in assessing the innovation capabilities of co-localized firms. Access to external tacit and codified knowledge depends on the extent to which effective communication among innovators takes place through the innovation system. In this context, the properties of regional economic systems, conceived as communication networks into which information flows, matter in explaining the capability to generate new technological knowledge.

The remainder of this chapter is organized as follows. In Section 1, the basic elements of the economics of technological communication are laid down. In Section 2, a simple formal model of the interaction of the stylized facts is presented. Section 3 summarizes the evidence about the channels and flows of technological communication within technological districts and the main findings are reviewed and structured. In Section 4, the distinction between increasing returns and the dynamics of positive feedback is outlined. Section 5 summarizes the main findings.

1. TOWARDS AN ECONOMICS OF TECHNOLOGICAL COMMUNICATION

The notion of technological communication makes it possible to appreciate the role of technological externalities and yet complement it with the notion of transaction costs in the absorption and communication of external technological knowledge. While the notion of technological externalities is consistent with the Arrovian notion of technological information—a public good with low levels of appropriability and excludability—it omits the key role of the specific costs that decodification and understanding of the available complementary knowledge and information entail. The traditional approach in fact assumes that technological externalities do spill freely into the environment and no provision is made to take into account the relevant costs of search, decodification, and assessment of existing technological knowledge dispersed among a myriad of agents and buried in tacit and idiosyncratic procedures. Technological communication differs from technological externalities. Too much emphasis has been put in the innovation systems literature on technological externalities as if external technological knowledge could be acquired freely in the 'atmosphere' without dedicated efforts (Hayek, 1945; Lamberton, 1971, 1996).

Knowledge externalities are mainly local and specific, as opposed to global and generic. Technological communication is relevant in the former case. Dedicated activities must be put in place for firms to be able actually to take advantage of external knowledge. Localized technological knowledge spilling into the atmosphere is highly specific and idiosyncratic. As such, it can be understood only by a limited subset of agents; it can be assimilated usefully only when some complementary conditions hold and it can apply only to a limited set of techniques, markets, and business circumstances. All this reduces the scope for spontaneous and free application. In other words, it is not sufficient that technological externalities are freely available in the air for firms to make effective use of them as intermediary inputs in the production of new technological knowledge. Substantial communication costs are to be accounted for and the context in which technological communication can take place needs to be qualified and the key factors identified. The notion of technological communication seems far more appropriate to the new theorizing about the quasi-private nature of localized technological knowledge from the allocation viewpoint. It seems also appropriate to accommodate the view that technological knowledge has strong elements of a collective good from the generation viewpoint.

Because many knowledge externalities appear to be local, instead of global, the notion of communication costs becomes relevant. Knowledge externalities can be treated as a signal which can be heard, understood, and recombined only within a given space, defined in terms of proximity and homogeneity. Dissipation of the signal increases with distance. For a given strength of the signal, the length of the dissipation ray which relates the intensity of the signal to distance, defines the size of the regional and technological systems within which externalities are accessible: the signal can be heard.

Standard communication economics should pay equal attention to both the emission and the reception of signals. In the economics of technological communication, however, it may be argued that reception matters more than emission. Local technological externalities are generated by innovators and other learning institutions without any specific and dedicated activity: technological knowledge spills over. The economics of technological communication hence focuses all the activities that lead to improve the capability of agents to identify, listen, understand, and reutilize the messages that flow in a limited portion of the 'atmosphere'. Technological externalities are relevant as long as they can be identified, understood, and eventually assimilated. The economics of technological communication in fact discriminates between global and local externalities and defines the access conditions to external technological knowledge.

Technological communication requires specific resources in order to take place. Proper inputs are necessary in order to locate the signals and channel and convert them into messages that can be understood. According to growing evidence in communication studies, effective communication can take place only when communication channels are put in place. Once a communication channel has been established, the amount of traffic, that is, the flow of signals and messages, can vary widely, as can their quality. In order for signals to be understood and for receptivity to take place, common protocols and communication codes have to be established. Hence the relevant object of analysis in understanding communication is the architecture of the networks.

The efficiency and structure of the communication network become a central issue here. The efficiency of the communication system in place within a network depends on many factors such as the use of the same communication codes, effective interfaces and communication protocols, low levels of redundancy, and dispersion of signals. Efficiency is also affected by the structure and architecture of the communication network.

A distinction can be made between communication networks according to their structure: geodesic networks, structured networks, oligarchic networks, and organized networks. In a geodesic network, all agents are directly connected to each other: the total number of links that are necessary to connect members of a network to each other increases exponentially. The entry of one additional agent in the network implies the creation of as many additional links as there are agents in the network. With given unit communication costs, total communication costs increase exponentially.

In a structured network, on the other hand, interconnection is viable and additional links can take advantage of existing ones. The entry of a new agent in the network leads to the building of just one additional link: the new agent can take advantage of the existing links between all the other agents. Unit costs in technological communication here are characterized by economies of scale and density. Investments in communication capacities, both in terms of dedicated skills and social norms, are long-lasting and have a large capacity: hence relevant economies of density are at play. Over time, average communication costs, because of the effects of economies of density, exhibit a declining slope. Moreover, if additional channels are added to existing ones, their costs are incremental rather than marginal. Within structured communication networks, externalities in the communication of technological knowledge are very strong when all parties engaged in technological communication 'share the same code' and hence benefit from a radical reduction in the huge decodification and search costs necessary to locate the bits of existing and external knowledge which can be directly relevant to each firm's generation process of new technological knowledge. Technological communication entails significant externalities when indirect communication flows are taken into account. When agents are distributed in an economic space along a sequential line and the communication flow is unidirectional, it is easily seen that communication between C and B benefits from communication between A and B. Triangular communication is especially important when A and B do not communicate directly but only via C. Triangular communication can have important consequences, both within industries when C is a supplier or a customer of both A and B in the same industry, and between industries when technological information flowing vertically in the industrial output matrix between A in industry 1 and B in industry 2 is relevant for agent C in the lateral industry 3. Within structured networks, technological communication itself is a collective good where each agent and each party is interested in enhancing the communication conditions among all members of the community. Technological communication, moreover, is an interactive process where both parties are actively and purposely involved in a barter relationship.[1]

Within oligarchic networks, some firms are able to retain proprietary control over the traffic flow so as to become stars, that is 'switches'. Switches are able to filter the flow of communication and take advantage of their central position. Here the location of each agent in the architecture of the network becomes extremely important. At any point in time, we can observe at each layer well-defined structures of communication flows, where some agents are better located than others in that some agents have access to more communication links than others and some agents happen to have access to more effective links than others. Specifically, we see that some agents can be more receptive than others to the given level of technological externalities available. And we also see that some agents have access to technological externalities at low communication costs, while others do not have such access at all.

Finally, in organized networks, an actual division of labour takes place. The emergence of firms specializing in the provision of technological information to the rest of the system is the distinctive issue. The provision of knowledge-intensive business services on a quasi-competitive basis makes it possible for all the firms in the network to access the flow of technological information and external knowledge. In this case, knowledge-intensive business-service firms act as open nodes in the network. The structure of the market-place for such services becomes a key factor in the accumulation of new technological knowledge: the closer it is to competitive conditions, the better it is for the rest of the system.

It is now clear that the levels of effective technological communication depend not only upon the resources devoted by each agent to establishing technological connections with other firms and academic and scientific institutions within innovation systems nor upon the amount of information that each firm is able to receive and actually assimilate from the innovation system in which it operates. It is also and mainly conditional upon the structure, the architecture, and the quality of the network in which each firm is located.

---

[1] See the analysis of know-how trading as a barter relation by Carter (1989).

Communication costs, in fact, depend upon the heterogeneity of actors, their distribution in both the regional and technological spaces, and the variety of communication codes in place.

The dynamics of communication costs and local externalities provides the basic engine to understand the evolution of technological districts.

2. A SIMPLE GEOMETRIC EXPOSITION

A simple geometric exposition of our approach can be based upon the seminal contribution by Arrow (1962a). Appropriability conditions are the basic factor accounting for the divergence between private and social output in the generation of technological knowledge. Each firm invests in the generation of new knowledge a smaller amount than socially required because of the reduction in revenue associated with uncontrolled imitation. When both the positive and negative effects of external knowledge and the conditions for informal know-how exchange are qualified, however, a different picture emerges: one where the amount of knowledge produced is larger, the larger the number of firms able to exchange know-how, where the divergence between private and social output is lower, and where a collective output can be identified. For a given number of firms engaged in complementary innovative and communication activities, the collective technological knowledge output is now higher as a result of the positive effects of external economies. The distance between the levels of the knowledge output function with externalities and without externalities measures the effects of external increasing returns in the production of new knowledge.

External technological knowledge, however, does not fall from heaven like manna: it is now an input, which can be quasi-internalized with specific absorption and 'listening' costs which depend upon the efficiency of the communication system. The latter, in turn, is affected by the variety of codes and the number of communication channels. The costs of the production of knowledge, including such communication costs, are lower for firms that are able to establish cooperative relations and access the pool of collective knowledge now made available. Appropriability is also affected. The opportunity costs engendered by the uncontrolled leakage of technological knowledge are lower, the higher are the mutuality and trust conditions in place within the group of firms.

With given research costs, optimum levels of knowledge production can be identified: between the private and socially desired optima, a collective optimum can now be easily identified. It stands in-between the Arrovian private and social optima and makes it possible to reduce the social loss due to inappropriability. Most important, the collective optimum makes it possible to appropriate many benefits of external increasing returns in the production of knowledge: the larger the number of firms, the larger the amount of knowledge generated. In the long term, however, such a positive relationship takes place with diminishing returns associated with exhaustion of the local and modular complementarities among technological knowledges.

On this basis we can recall equations (3.1) and (3.2) from Chapter 3:

$$LTK_i = j(R\&D\&L_i, EK_{n-i}) \qquad (6.1)$$

where

$$EK_{n-i} = m(N); \quad m' > 0, \text{ but } m'' < 0, \qquad (6.2)$$

A closer inspection of the cost equation associated with the generation of knowledge is now necessary. In order to make good use of the potential knowledge externalities available, prospective users have to undertake specific communication activities and bear the related communication costs which add on to traditional research activities.

Let us assume a given distribution of agents in the regional, technological, and economic spaces. According to the localized technological knowledge approach, each agent differs from the others and has some idiosyncratic features. Each agent undertakes some innovative activity and spills locally some technological externalities whose accessibility decays with distance and the heterogeneity of agents. Each agent tries to take advantage of the local externalities made available by the other firms engaging in communication activities which become more and more expensive as distance increases. Distance here is measured in terms of the cumulated differences among agents with respect to the regional, technological, and economic spaces: the larger the number of firms, the larger their cumulated distance index.

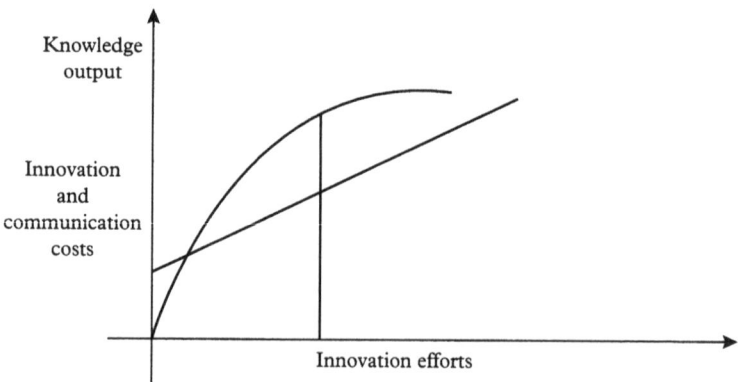

Figure 6.1. *Short-term conditions in the production of localized and collective technological knowledge at the firm level*

Innovating firms bear innovation and communication costs, consisting of the costs of research and learning inputs and of communication inputs. In the short term, the former can be stylized as variable costs, while the latter are fixed ones. The former vary with their quantity. The latter, by contrast, increase only with the number of connected firms and communication channels which can change only in the long term because of the entry of new firms. The dynamics of fixed costs, in the short term, apply: a communication channel, once established, can carry large and increasing quantities of communication flows (Arrow, 1974 and Antonelli, 1999a).

In the short term, this assumption has relevant consequences. Firms bear the variable costs of internal research and development and learning activities (R & D & L) and the fixed costs of existing communication channels. A short-term equilibrium condition can be identified. External increasing returns in the production of knowledge are the result of two classes of factors: knowledge externalities as well as economies of density and incremental costs in communication. Firms, in fact, maximize the amount of internal R & D & L expenses under the control of the total revenue stemming from innovating activities, for given levels of communication channels. R & D & L expenses include some fixed communication costs. Because of the knowledge externalities spilling from the other connected members of the network, insiders earn extra profits.

Connected firms take advantage of increasing returns and this privilege attracts new firms. Learning to communicate spreads and

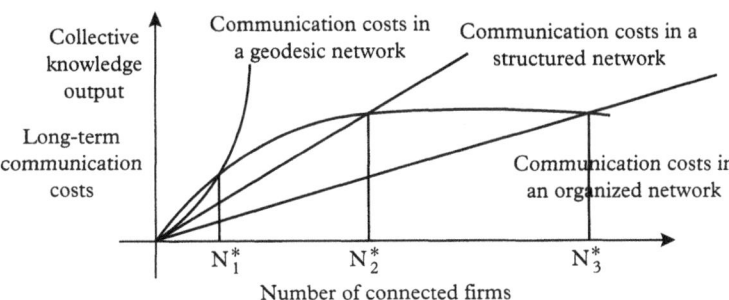

Figure 6.2. *Communication structure and collective knowledge output in a technological district*

the entry of new agents within the network drives the system towards long-term equilibrium.

In such a long-term framework, the equilibrium size of each district will depend upon the trade-off between the net positive effects generated by the addition of each agent to the network on the productivity of innovative activities, after taking into account the loss in terms of appropriability, and the negative effects of the additional communication costs. It is immediately clear that the lower are the communication costs among agents and the better the quality of communication channels, the larger is the number of firms and the larger the size of the district, and, most important, the larger is the scope for increasing returns available in the production of technological knowledge (see Fig. 6.2).

Within a network, the number of channels or links (L) increases with the number of connected agents (N). In a geodesic network, the number of links increases exponentially with the number of connected agents. Hence we have that the maximum number of links is:

$$L = N(N-1)/2 \tag{6.3}$$

The structure and efficiency of the network becomes a central issue here. All increases in the efficiency of the communication system in terms of new enhanced communication protocols, more effective interfaces, reduction in redundancy and dispersion increase the amount of collective knowledge which can be generated by the network.

The important distinction between geodesic networks, open-interconnection networks, oligarchic networks, and organized networks has to be stressed again. Within geodesic networks, all agents are directly connected to each other. In such networks in the long term, sharp dis-economies of scale take place. Within structured networks, based upon open interconnection, communication costs are characterized by economies of scale and density. Unit communication costs, in fact, decline with the number of links because additional channels can take advantage of the dynamics of incremental costs. Within oligarchic networks, the location of each agent with respect to the traffic flow and the architecture of the network plays a central role. The place in the network is central to assess the actual access to potential knowledge externalities and its communication costs: some agents can be better placed than others. In organized networks, characterized by high levels of division of labour and the specialization of firms in the provision of technological communication, unit communication costs exhibit constant returns to scale.

Formally we see that, at the aggregate level and in the long term, for given levels of research and development activities, firms will engage in technological communication (TC) and enter the connected system as long as profits are driven to average levels (zero in perfect competition). Specifically, on the revenue side, we account for the effects of external knowledge on the production of new localized and collective technological knowledge. These effects can be thought to be positive but with a diminishing impact as the number of firms and the related variety of knowledge spillovers increase: complementarity is limited. On the cost side, we find the communication costs (CC) that are necessary to establish communication links.

Here a major divide takes place between structured and geodesic networks. In the latter, we can assume that total communication costs increase with a more than proportionate effect with respect to the number of agents (N) engaged in the process. In the former, total communication costs increase with a proportionate effect. In both cases, the equilibrium size of the network is reached when total communication costs cross the positive effects of knowledge externalities. Hence:

$$P(TC) \text{ is max for } LTK(N) = CC(N) \qquad (6.4)$$

Firms learn to communicate, engage in technological communication, and benefit from external increasing returns until the maximum number of firms and maximum levels of heterogeneity among firms are achieved. Firms will benefit from external increasing returns as long as the equilibrium size is achieved. By the same token, firms will build new communication channels and establish additional communication flows until the equilibrium size is achieved. Until such size is achieved, clearly both disequilibrium and increasing returns coexist. In the long term, not only do firms learn about opportunities to benefit from the increasing returns associated with access to local knowledge externalities, but they can also learn about the design and governance of the network so as to be able to achieve an open-interconnection structure.

Analysis of the conditions and context for effective technological communication to take place and possibly to improve (and this is most important) the structure and governance of the communication network, lies at the heart of the innovation system approach and provides the basic conditions for localized technological knowledge to become collective and for external increasing returns in the production of knowledge to take place.

Within communication networks, we see in fact that at each point in time, the magnitude and impact of the effective flow of information which is both emitted and received by each agent can be thought to be the outcome of the interaction between two classes of stochastic events: (1) the connectivity probability that the flows of effective communication and the exchange of information take place; (2) the receptivity probability that the results of the research and learning efforts of each firm in the system are effectively assimilated. In this context, however, the structure, architecture, and the general efficiency of the network, together with its institutional design, play a major role as they are the main determinants of the general levels of communication costs (David and Foray, 1994; Krugman, 1996; Antonelli, 1999*a*).

According to the hypotheses outlined above, it is assumed that the production of technological knowledge by each firm can be formalized as a result of the interaction of internal research and learning activities and the creative and intentional access to external technological knowledge, characterized by significant complementarity, and its actual implementation. The complementarity of the different technological knowledges is an important feature for a local innovation system to become an effective innovation environment. Heterogeneous pieces of technological knowledge which belong to different modules of technological knowledge have little chance to affect each other even with high-quality communication systems in place.

The different levels of effective communication among innovators, as measured by the mixed probability of communication processes, are likely significantly to affect the productivity of the total amount of resources devoted by each firm to research and learning activities and hence to reduce innovation costs substantially (Nelson, 1987). This, in turn, helps increase the absolute levels of innovation introduced within the local innovation system.

This process is especially evident within technological districts such as Turin in Piedmont, Modena and Bologna in central Italy, Toulouse in France, Route 128 and Silicon Valley in the USA[2] (Antonelli, 1986a; Russo, 1985, 1996, 2000; Dorfman, 1983).

Because of their stochastic nature, however, communication processes may also fail. Localized technological knowledge, because of its idiosyncratic character, is appropriated by innovators and is no longer collective; increasing returns are no longer available. The general efficiency of innovation activities declines. Firms fail to generate real total-factor-productivity-enhancing innovations and fall into the irreversibility trap. When technological communication fails, the generation of new knowledge and the related introduction of technological change become more expensive. Firms with a large endowment of superfixed production factors can face increases in demand and changes in the relative prices of variable production factors only with a substantial decline in technical efficiency and limited rates of introduction of technological innovations. Agglomerations decline and industrial districts decay. This is also the story of many traditional industrial districts in central England and northern France. The conditions for technological communication become a key issue of central relevance for the economics of innovation and new technology, as well as for regional economics.

---

[2] The stochastic nature of communication processes, however, makes such self-reinforcing feedback mechanisms random. Mixed communication probabilities are especially sensitive to all perturbations in both connectivity and receptivity probabilities. In such conditions, local innovation systems may eventually experience a sharp reduction in general communication efficiency and reverse negative feedbacks may take place with major discontinuities in long-term growth patterns.

3. TECHNOLOGICAL COMMUNICATION AND INNOVATION WITHIN TECHNOLOGICAL DISTRICTS: A REVIEW OF THE EVIDENCE AND AN AGENDA FOR EMPIRICAL RESEARCH

Location plays a major role in favouring technological communication, but agglomeration *per se* is not sufficient for technological communication to take place. Technological communication takes place at a variety of levels. Significant scope for empirical analysis emerges when different forms of interactions of firms in different market-places are analysed from the viewpoint of their implications for communication. This analysis is especially relevant when it takes into account the variety of channels by means of which technological communication can take place within technological districts as well as the variety of positions within each network of each agent.

A tentative review of the rich and still growing empirical literature makes it possible to list eight different and relevant communication channels.

(1) *Local labour market conditions* Labour markets provide important opportunities for technological communication to take place. Interfirm mobility greatly enhances the dissemination of information. As a matter of fact, external labour mobility is a basic factor in the recombination of existing information and in the regeneration of a common information pool within an economic system. Intrafirm labour mobility performs a similar role at the company level, although within a narrower field of action. Hence we can expect that the greater interfirm labour mobility, the greater rates of technological communication. Economic systems where seniority within companies plays a major role can suffer from a reduced level of technological communication. Even worse can be the case of an economic system with reduced interfirm and intrafirm labour mobility. It is clear, in fact, that if labour mobility that is too rapid can reduce the scope for learning processes to take place, a rigid allocation of personnel to limited tasks within the same company can impede the dissemination and recombination of all technological information (Brusco, 1992; Edquist, 1997; Salais and Storper, 1992; Clarysse *et al.*, 1995).

(2) *Advanced local financial markets* Information asymmetries in financial markets lead to credit-rationing and are a major source

of advantage for large corporations over small companies and start-ups. New companies face problems finding both credit and risk capital because of the difficulties of assessing the commercial prospects of new products and new processes. Large corporations with internal capital markets can better assess the market profitability of new ventures. Here proximity and repeated interactions between lenders and prospective shareholders and new and small companies can play a major role in reducing credit-rationing as well as credit rates. A local advanced financial community can better provide access to the technological information and technological knowledge that are necessary to screen new undertakings and assess new projects both in terms of investments by small incumbents and creation of new firms. Proximity enhances trust and makes technological information more readily available. Both credit and risk capital can be provided within local financial markets under better conditions for small and new companies. Empirical evidence confirms that average credit rates for small firms and innovative projects are lower within industrial and technological districts than for stand-alone firms (Bellandi and Russo, 1994; Dei Ottati, 1996). The quality of local financial markets plays a role also in terms of well-known venture-capital effects with the provision of risk capital to new high-tech start-ups and eventual public offerings in the stock exchange. The transition of new private firms to the status of public companies, with shares traded on the stock exchange, is crucial for venture capitalist and risk-non-averse financial institutions to trade their assets and recover liquidity. Advanced local financial markets are also a conducive context for interorganizational mobility fed by mergers and acquisitions. An effective local financial community provides better conditions to practise the acquisition of new companies as a systematic source of technological knowledge and to change the borders of firms in terms of external growth via integration and diversification as well as spin-off of knowledge-intensive units with increased specialization. From a technological communication viewpoint, financial markets, especially for new innovative firms, can become a major tool to implement the 'mix and match' generation of new technological knowledge. Local accessibility to such financial markets in turn becomes a key factor in sustaining rates of introduction of technological change (see Scherer, 1999).

(3) *Local intermediary markets* The role of both upstream and downstream user–producer relations has been greatly appreciated

by much economic analysis. It seems clear that an economic system with an articulated industrial structure with many intermediary markets, where a variety of firms interact for the provision and purchase of specific intermediary inputs, can support technological communication much better than economic systems where vertically integrated firms control the full *filière* (Lundvall, 1985; Russo, 1986, 2000, Von Hippel, 1988; Langlois, 1992; Robertson and Langlois, 1995; Camagni, 1991, 1999; Saxenian, 1994).

The enhanced division of inter-industrial labour seems extremely relevant for technological communication to take place, as far as knowledge-intensive business services are concerned. Here, in fact, it is clear that vertical integration within manufacturing companies of an array of advanced services reduces the scope for dissemination and recombination of specific technological knowledge that, once generated for one single use, can easily be applied to a variety of different contexts. This approach makes it possible to appreciate the characteristics of regions in terms of sectoral composition and related opportunities for technological outsourcing of firms. The distribution and quality of knowledge-intensive business-service industries, in fact, have important effects on the economic system in terms of innovative capacity. An increase in the exchange of tacit knowledge, made possible by local supply of the services of consultants and advisers, improves connectivity between agents, sharing learning experiences and creating learning opportunities, and thus advances receptivity. Similarly, improved business services, in terms of distribution, capillarity, competence, and access, improves the interaction between tacit, localized knowledge and increasingly larger amounts of generic knowledge, and in so doing is conducive to the accelerated introduction of technological and organizational innovations and solutions specifically tailored to a firm's individual needs. An active local supply of knowledge-intensive business services can stimulate technological outsourcing and hence the demand for knowledge-intensive services by small and medium-sized firms in particular (Antonelli, 1999*a*; Russo, 2000).

Knowledge-intensive service firms activate important flows of triangular communication among firms and are often the result of the vertical disintegration of large companies with the spin-off of specialized service units within large corporations which eventually search for markets larger than intra-corporate ones.

Triangular communication along vertical *filières* plays a special role in this context. Firms active in the same industry rarely share technological information and participate in direct horizontal communication. The case for indirect communication, however, is often found when two firms in the same industry share a common supplier or customer. The common supplier or customer then becomes a node in a triangular flow of communication where two communication flows pass into a switching node.

(4) *The features of local industrial structures* Proximity and co-localization within a local system favour both intrasectoral and intersectoral dissemination of technological knowledge, both vertically and horizontally.[3] Intersectoral communication is especially relevant when general-purpose technologies are at play: agents are much less reluctant to share their knowledge with firms active in other markets. Relevant barriers to communication arise from the difference of codes and the idiosyncratic character of the information available. New technological knowledge generated in one industry, however, often has significant scope for direct application in other industries, either along common functions or along production *filières* even beyond user–producer interactions. As far as intersectoral dissemination is concerned, communication can be thought of as the outcome of a cooperative attitude of agents who can share new technological knowledge with little fear of harming appropriability conditions because of the difference of markets and customers. Again, indirect triangular communication enhances the probability of successful communication taking place when a supplier or a customer for firm A in industry 1 acts as a switching node for communication directed towards firm B in an industry 2 which is not vertically related to industry 1.

The reverse is true for intrasectoral communication: the risks for opportunistic behaviour are higher as well as the homogeneity of languages and codes. Technological communication can easily become a factor of imitation here. While collective innovation is harmed by appropriability concerns, the diffusion of both product

---

[3] An extensive literature refers to intersectoral externalities as 'Jacobs externalities' and contrasts them with MAR (Marshall–Arrow–Romer) intrasectoral externalities which characterize industrial districts (see Becattini, 1979, 1987, 1989; Brusco, 1992).

and process innovations is very fast. A trade-off between rates of innovation and rates of diffusion can arise.

The coexistence of large and small firms within technological districts appears to be a critical element for many reasons. Communication is enhanced by variety and diversity among agents. Opportunity for exchange of information between firms is greater, the greater is the difference between channels in terms of typology of knowledge accumulated, scope for its implementation, and sets of competencies. Small firms can benefit from faster decision-making and entrepreneurial reaction than large firms. The latter can access the advantages of economies of size in conducting research and development activities. Firms, according to size, differ with respect to the typology of knowledge they produce and the knowledge they use. Large firms have a clear advantage in the production of codified knowledge, while small firms excel in the accumulation of tacit knowledge. In turn, however, large firms mainly depend upon external tacit knowledge and small firms on external codified knowledge. The interactive coexistence of large and small firms is vital in enhancing the accumulation and circulation of technological knowledge.

Outsourcing plays an important role as a vector of communication flows, especially when large firms rely upon smaller suppliers within the same industry for the provision of specialized components which enter new products. Even stronger is the case within subcontracting relations where large firms retain control of the overall innovation process and induce major innovations, both in products, processes, and intermediary inputs and organization, in suppliers. Vertical relations become the appropriate context for important technological communication flows that are both formal and intentional and informal. Vertical relations among firms in this context are further enriched and implemented by the mobility of personnel from large to small firms. The implementation of new technologies conceived in large firms and actually manufactured by small firms makes the relationship even stronger.

Rich empirical evidence confirms that large firms act as '*de facto*' incubators. High-tech start-ups, both in manufacturing and more and more in knowledge-intensive business services, are often founded by former managers of local large corporations. Co-localization is important here as a way to secure and keep open business and personal relations with former employers. Their derived demand often represents a large share of sales in the early years.

The key role of large firms within technological districts marks an important difference with respect to the neo-Marshallian approach to industrial districts which emphasizes the role of small firms together with their homogeneity in terms of size (Antonelli, 1986a; Becattini, 1979, 1987; Camagni, 1991, 1999).

Industrial demography and, specifically, natality rates of new firms are one additional and important channel of technological communication. Newly founded firms, on the other hand, provide the opportunity for new ideas to be tested and hence communicated in the market-place. High-tech start-ups provide a basic contribution to the rate of technological advance. Rich and detailed empirical evidence is provided by Swann et al. (1998) who stress the differences among entrants and incumbents in terms of capability to absorb spillovers. Specifically they show that spillovers in radical innovations tend to induce entry rather than rapid growth among incumbents.

(5) *Interregional links* Technological communication between technological districts plays a key role. The stock of multi-regional companies localized within each technological district is an important vector of interregional communication. The larger the number of multi-regional firms and the larger the size of communication system of each of them, the larger the access of other co-localized firms to interregional communication flows is likely to be. Cross-entry of incumbents in other industries and other regions enhances technological communication because it provides opportunities for techniques and information well established in one industry and region to be disseminated to others. This is typically the positive effect of both the entry of external multinational companies and of the multinational growth of local companies. Multi-regional companies have an important opportunity to act as 'switches' among technological districts. In so doing, they can play a direct role in activating significant flows of triangular communication among technological districts connecting firms active in different contexts and yet complementary with respect to their technological competence and their innovation strategies. Such an opportunity is also an incentive. Multi-regional companies can, in fact, appropriate a large share of the positive effects of interregional links. The increasing internationalization of corporate R & D in a few selected

technological districts worldwide can be understood in this context as the result of the active internalization of the scope for technological communication and recombination of local pools of collective knowledge (Patel, 1995) provided by the selective multi-localization of research activities.

(6) *The features of the local innovation system in terms of knowledge infrastructure* The level of the local academic infrastructure, in terms of overall quality, breadth, and depth of the range of scientific activities, and the degree of interactivity with the local business community, plays a major role (Geuna, 1999). University and the academic community at large do spill scientific and technological externalities, as much empirical evidence has shown. Local interaction between universities and the business community, however, goes well beyond the traditional set-up. In the traditional organizational and institutional set-up, universities were expected to provide global externalities mainly through the publication of the results of their research and the training of postgraduate students, able to read the publications, who would enter the business community. Access to such knowledge externalities, however, appears viable only when (and if) it is complemented by actual access to local externalities. This, in turn, can take place only if and when academic and business communities have established clear ways of interaction and communication as is often the case in the US institutional set-up. In this context, not only is scientific excellence achieved in many fields, but the flow of postgraduate students from universities to firms is high. The funding of academic research activities by firms outsourcing extramural basic research activities is a third growing vector of technological communication between universities and the business community.

Two additional communication tools have emerged: academic entrepreneurship and academic supply of knowledge-intensive business services. Universities enter more and more the markets for knowledge-intensive business services and provide advanced services, especially to local companies with a clear bilateral advantage. Academic researchers are exposed to specific technological problems with beneficial effects on ongoing scientific research. Local companies can access a pool of advanced competencies and a dedicated infrastructure, often characterized by strong indivisibilities and huge unit costs, which are provided at low incremental costs. Co-localization here favours repeated interactions in the market-place and their implementation through personal contacts.

Academic entrepreneurship is more and more becoming the fifth major channel of technological communication between universities and the business community. Universities become incubators of scientific entrepreneurs on many counts. Researchers are professionally solicited to exploit directly new scientific breakthroughs by elaborating technological applications: the creation of start-ups is welcomed in the new academic ethic. Secondly, universities are actively involved in supporting academic start-ups with the provision of infrastructure and space: proximity plays a major role in sustaining academic start-ups. Start-ups are founded in the region nearby as a way to minimize risks. Proximity here is an insurance against failure: exit from university is often delayed and can be resolved only when and if the new venture is successful.

Academic entrepreneurship, extramural research activities conducted on behalf of firms externalizing scientific activities, and academic supply of knowledge-intensive business services enhance the local character of knowledge externalities spilling over from universities; while the traditional tools—academic publications and the supply of graduate trainees—has a much lower local content and rather a global reach. The interaction between global and local externalities here becomes crucial in favouring the effective capabilities of technological districts endowed with a strong academic community to sustain high rates of accumulation of technological knowledge. This is rarely the case in the European institutional set-up, where as a matter of fact the empirical evidence upon the positive effects of academic externalities is much less strong. In continental Europe, however, the strong commitment of academics in private consulting, especially when favoured with part-time practice, seems conducive to the rapid dissemination of new technological knowledge. A trade-off between dissemination, favoured by private consulting, and generation, implemented through institutional participation, seems to emerge in this context (Mansfield, 1991; Bania et al., 1993; Jaffe et al., 1993; Audretsch and Stephan, 1996; Feldman and Audretsch, 1999; Russo, 2000).

Localized interaction between education and training and production are all the more useful for enhancing rates of technological communication within a system. Local supply of specific training capabilities can favour greatly the external learning process due to the higher content of idiosyncratic information and competence that trainees can accumulate and the stronger compatibility of their generic knowledge base with the localized character of the technological knowledge used in each specific district. The alternance of training spells throughout the productive life of individuals can greatly favour the circulation of information within an economic system. The traditional concentration of education in youth with limited access to retraining later on in life clearly reduces the opportunity for technological communication to take place. On the other hand, permanent training with a strong local content can assist technological communication because it makes it possible to generalize the specific acquisitions of workers and teachers in previous jobs (Edquist, 1997; Freeman, 1991, 1997).

The regional concentration of the research laboratories of industrial firms adds to the opportunities for smaller firms located in the vicinity to take advantage of technological externalities at low communication costs and enhances the probability that firms can take advantage of interstitial technological opportunities that are considered internally as second best by larger corporations and yet can lead to profitable technological innovations for smaller firms. The regional concentration of research laboratories can also become the institutional device for symmetric communication externalities to take place among large firms in a context of increased division of labour based on mutual outsourcing (Howells, 1990; Quéré, 1994; Patel, 1995).

Regional concentration of research laboratories and headquarters of large companies favours technological entrepreneurship, whereby new start-ups and the natality of new high-tech firms are the result of the exit of former researchers from large R & D centres. Former researchers of large corporations have, in fact, a clear incentive to locate the new venture in strict proximity to their former activities in order to minimize costs and to take advantage of ongoing relations with former colleagues.

New emerging markets for disembodied technological knowledge where firms sell and buy patents, know-how, and trade technological licences seem to pave the way to increasing specialization of regions and firms in the generation of dedicated technological knowledge. The active search for licences of patents, and know-how, can help firms access external knowledge available on international and domestic markets. Such external technological knowledge can be recombined and contribute to the internal elaboration of tacit and codified knowledge, with evident advantages in terms of the efficiency of intramural R & D activities (Arora and Gambardella, 1990; Arora, 1995, 1997).

Local markets here can be far more effective than global ones on two counts. Co-localization actually reduces the risks of opportunistic behaviour of licencees: control and retaliation are easier. Co-localization makes it easier for vendors to better control not only the current use of their technologies, but also to retain some ability to appropriate the advantages of learning by using and learning by doing which might eventually emerge in the licencees' premises. Co-localization also reduces the significant post-sales costs in the adaptation of technologies conceived elsewhere. User–producer relations are also made easier by proximity in this context.

The acquisition of start-ups is often a way, for larger corporations, to trade technological knowledge embodied in firms. Co-localization favours their integration and coordination within the organization of acquiring firms.

(7) *The quality of local communication infrastructure* Emphasis on the role of technological communication makes it possible to appreciate how the characteristics of the present wave of innovations in communication technology, itself a product of the clustering of localized and complementary technological changes, are likely to interact with the rate of introduction of localized technological changes and to enhance the general levels of innovation capability of firms. The quality of local communication networks can play an important role in favouring the division of innovative labour when high-speed data communication can take place and high-definition images can easily be transferred among research units. As growing evidence confirms, digital communication is a complement rather than a substitute for person-to-person communication. Technological districts with high-quality communication infrastructure can benefit from the spiralling interactions between digital and face-to-face communication (Antonelli, 1999a).

Secondly and most important, a large amount of empirical evidence shows that urban and especially metropolitan areas provide a very positive environment for communication to take place and hence more opportunities to foster the rate of technological change (Castells, 1989). Proximity and spatial density enhance technological communication informally also as a result of the higher probability of repeated interactions among heterogeneous and yet complementary agents. Co-localized firms within metropolitan areas have more opportunities to share a common language and hence to save on the costs of codification and decodification of information about technology as well as business conditions (Allen, 1983; Freeman, 1991; Utterback, 1994; Harrison et al., 1996).

This is the third and relevant difference between Marshallian industrial districts and technological districts.[4] The traditional Marshallian district, in fact, especially in the Italian literature, is mainly characterized as a regional space with lower levels of population density and low levels of intra-regional concentration of plants and firms. Metropolitan areas seem better able to provide the mix of variety and complementarity of economic activities, endowment of scientific infrastructure, and high quality of communication systems which favour technological communication.

(8) *Localization and technological strategies* Location plays a major role in favouring *ex-ante* coordination among innovating agents. As Richardson (1972) had clearly anticipated, *ex-ante* coordination plays a key role in the 'transfer, exchange or pooling of technology' (p. 892). According to Richardson,

the essence of co-operative arrangements... would seem to be the fact that the parties accept some degree of obligation—and therefore give some degree of assurance—with respect to their future conduct. But there is certainly room for infinite variation in the scope of such assurances and in the degree of formality with which they are expressed. (p. 886)

Better information on the research agenda and mutual understanding of the competencies of agents are available within technological districts: this favours the division of innovative labour enhancing specialization in complementary but dissimilar innovation activities and reduces duplications. Opportunistic behaviour is also restrained within technological districts because of the long-term interactions associated with co-localization: co-localization, as a matter of fact, can be thought of as a symmetric hostage. Transactions in disembodied technology and licensing agreements may be

---

[4] See Becattini (1979, 1987, 1989); Bellandi (1989); Bellandi and Russo (1994); Brusco (1992); Russo (1985, 1996).

easier for firms co-localized within technological districts: continuity in the relationship favours the supply of technical assistance. As a consequence, better implicit coordination of investment decisions is achieved within technological districts where information about market conduct and the technological strategies of each agent is easier to gather.

Professional associations and industrial clubs provide important opportunities for technological communication to take place and should be considered key factors in the definition of the organization of an industry. Ever since the path-breaking analysis of Richardon (1972), professional associations, including collective research institutions, are seen as basic institutions that facilitate the diffusion of relevant knowledge within limited regional spaces and are conducive to a variety of forms of tacit exchange of information and know-how. Locally, technological cooperation is often the result of implicit strategic actions taken by co-localized firms to increase connectivity and receptivity levels and hence technological communication (Dorfman, 1983; Saxenian, 1994; Clarysse *et al.*, 1995; Hagerdoorn, 1995; Russo, 2000).

To a large extent intentional co-localization within technological districts and active participation in local communication systems can be thought of as a distinctive form of technological cooperation which can complement and even substitute for more formal technological partnership within footloose technological clubs.[5] The quality of receptivity and connectivity among agents can be influenced by intentional strategies such as location in close vicinity to other innovators. Specifically, empirical evidence suggests that the size of firms matters in this context. Large global corporations can actually substitute intentional co-localization for formal technological cooperation: localization in technological districts where highly idiosyncratic collective knowledge is available is common practice by large companies, with the location of research units and even pilot plants in technological districts and in the vicinity of advanced academic centres. Co-localization, on the other hand, is often the basis for more explicit and formal technological cooperation among small firms. Proximity offers the opportunity to implement latent complementarities and take advantage of them.

---

[5] See Fransman (1995, 1999) for detailed analyses of different forms of cooperative organization in innovation activities.

Localization within technological districts, moreover, is also relevant for reasons other than communication as it can help *ex-ante* cooperation to take place, favouring the coordination of different agents both with respect to technological strategies and investment decisions (Richardson, 1960).

4. INCREASING RETURNS: EXTERNALITIES AND POSITIVE FEEDBACK

The rich empirical evidence now available on technological districts makes it possible to introduce an important distinction between increasing returns stemming from externalities-cum-technological communication and the dynamics of positive feedback. The output of innovation activities increases locally, when and if the conditions for collective access to pools of technological knowledge apply, with a more than proportionate effect, not only because of the traditional externality argument, but also due to dynamic positive feedback (David *et al.*, 1998).

Externalities lead to increasing returns because of the indivisibility of technological knowledge. Each addition to a common pool of a bit of knowledge kept separate until then, because it had been appropriated by intellectual property rights or covered by confidentiality, has a positive effect on the general efficiency of the activities that generate technological knowledge, when and if a context conducive to technological communication is found.

Positive feedbacks lead to increasing returns effects for a different reason. For given levels of irreversibility, firms react by introducing larger amounts of technological changes according to the 'efficiency' of innovation activities. The higher the efficiency of innovation activities, the larger is the amount of innovation activities actually performed. Higher levels of innovation activities, made possible by good technological communication conditions, are likely to increase the amount of both internal and external stocks of knowledge available and lead to ongoing research and learning activities, with clear effects in terms of both economies of density and knowledge externalities. This, in turn, affects positively the efficiency of research activities and further stimulates the innovative efforts of firms. At each point in time, in fact, firms facing the irreversibility trap are more likely to innovate and hence to increase the amount of funds invested in research, development, and learning activities. This, in turn, increases the amount of research activities going on within the technological district and the local pool of collective knowledge becomes larger. This makes it possible to benefit further from increasing returns in the production of localized technological knowledge stemming from external knowledge economies, provided appropriate communication systems are in place. In turn, local communication conditions at time t are likely to affect the behaviour of agents not only with respect to the levels of their innovation activities but also to the levels of deliberate action taken to build up connections and receptivity which can enhance the efficiency of funds invested in research activities. All the characteristics of a self-reinforcing mechanism, based upon three-pronged positive feedback, are now in place.

When communication externalities are added to the dynamics of positive feedback, a generalized 'Matthew effect' is likely to take place (David, 1994; David et al., 1998).

As should now be clear, within technological districts increasing returns take place as a result of self-reinforcing interaction between three factors: (1) the dynamics of positive feedbacks in the definition of viable levels of innovation activities which can be conducted; (2) the amount of communication efforts which are being implemented, and (3) their efficiency as determined by access to local knowledge externalities.

5. CONCLUSIONS

The approach elaborated so far has important economic implications. Within atomistic competition the production of technological knowledge is doomed to be suboptimal because of limited appropriability and hence insufficient incentives for producers to commit resources to such a risky activity. Private and social optima diverge. Within technological districts, characterized as networks of connected innovators, knowledge acquires the attributes of a collective activity: it is no longer just an output but also an input in further activities. The larger the amount of collective knowledge, the smaller the divergence between private and social optima. It is clear, in fact, that total factor productivity levels at each point in time are determined by the amount of collective knowledge available within a system. The latter is clearly influenced by the size of technological districts. Hence we see that the larger the number of technological districts within an economic system, the faster is the rate of growth of the number of connected firms within each technological district and the larger is equilibrium level of the size of each district, the larger are the general levels of efficiency of the economic system.

The identification of technological districts becomes a central issue. Regions can be classified according to at least four parameters. The actual complementarity of the diverse and fragmented bits of technological knowledge implemented by each firm within a region is the first main factor. The absolute levels of research activities, as measured in terms of inputs, is clearly a second main factor together with the density of knowledge-intensive activities with respect to the number of firms and their output. Finally, the quality of the communication system in place is the fourth main parameter. The efficiency of research activities and the relationship between input and output is clearly affected by both complementarity and interactivity among firms.

For given levels of technological complementarity, levels, and densities of knowledge-intensive activities, all advances in the quality of communication systems are likely to have positive effects on the size and efficiency of each technological district. All reductions in communication costs make it possible to increase the size of districts and, most important, the amount of technological knowledge which can be produced in efficient conditions. Clearly, telecommunications are an important enabling technology for technological communication to take place: the diffusion of new telecommunications and communication technologies is likely to reduce the costs of technological communication and hence to increase the size of technological districts. All policy interventions which identify technological communication as an important goal can contribute to the reduction of these specific costs and again increase the size of technological districts and, relatedly, the amount of collective knowledge.

Secondly, and most important, all advances in the organization of the communication network make it possible to reduce the number of necessary links and consequently to reduce the negative effects of the exponential increase in total communication costs. The evolution from a geodesic network into more structured network designs, where interconnection is viable and communication can take place through pre-existing links and channels, has the evident effect of increasing the number of firms which can participate efficiently in the network and the amount of collective knowledge a system can generate. Learning to organize the communication network becomes a central issue and provides considerable scope for policy intervention, especially at the local level.

The interaction of the different channels through which communication operates and the quality of each communication system, as well as the appreciation of the role of each agent within each communication system, provide the final picture which can approximate the actual capability of firms to participate in the general communication process and the opportunities firms have to take advantage of existing and yet fragmented pieces of technological knowledge available in the system. The innovative behaviour of firms is deeply affected by the local 'milieu', in that this provides the communication context and hence affects the interplay of classical agglomeration effects with specific firm effects and the features and characteristics of the distribution of existing technological knowledge, now viewed as an essential intermediary product. The complexity and multi-layered dimensions of local communication systems provide an important agenda for empirical research. Communication channels actually complement each other, especially within technological districts where proximity and agglomeration provide a complex web of interaction mechanisms.

In turn, regional clusters can be ordered in terms of the variety and complementarity of communication channels in place: technological districts can now be defined as regions where knowledge externalities and low communication costs are especially conducive to making localized technological knowledge collective. These conditions, in turn, make possible increasing returns in the production of technological knowledge and add to the positive effects of agglomeration in terms of pecuniary and technical externalities with a generalized 'Matthew effect'. Enhanced efficiency in innovation activities and actual increasing returns fuelled by positive feedback are central in fostering the rate of introduction of technological changes.

Firms located within technological districts have more chance of being able to confront irreversibility costs and the irreversibility trap with the introduction of localized technological change. Such localized technological innovations are likely to exhibit high levels of total-factor-productivity growth as well as technological continuity and complementarity, both with respect to their existing capital stocks and to the innovations of co-localized fellow firms.

New evidence about the strong regional concentration of innovation activities and specifically about technological districts suggests that the so-called tragedy of the commons needs to be revisited. Collective property is the result of failure to assign property rights and leads to excessive use of common resources and overutilization of given resources. The definition of property rights over indivisible resources can, however, generate other inefficiencies in terms of rationing (Stiglitz, 1994).

The notion of collective knowledge makes it possible to reconsider the tragedies of the commons. When collective knowledge matters, the access of each agent has a twin effect: on the one hand, it does imply the use of common resources, but on the other hand, it helps increase the common pool. The external cumulability of technological knowledge, based upon the complementarity of the knowledge possessed by each learning agent within a technological district, explains the dynamics of the collective accumulation of technological knowledge.

Within a technological district, technological communication takes place in a context characterized by a complex mix of spontaneous and yet controlled forms of social interactions. Exchange of information takes place both formally and informally in quasi-markets which complement and assist the actual trade of goods and services. Firms accept some leakage in their proprietary knowledge, provided adequate returns are made available.

Analysis of the conditions for technological communication and the context for technological communication lies at the heart of the innovation system approach and provides the basic conditions for localized technological knowledge to become collective and for external increasing returns in the production of knowledge to take place.

Location plays a major role in enhancing technological communication due to its positive effects on both connectivity and receptivity effects, provided a variety of communication channels are in place. Each of these local communication channels differs widely in terms of effects, because of the role of density and interactivity. Moreover, agents may have a differentiated role within each communication system. Agents can be marginal or central, they can be well connected or poorly connected. Agents can be poorly connected to strong actors and well connected to weak nodes.[6] Opportunistic behaviour, however, is restrained within technological districts as a result of the long-term interactions associated with co-localization: co-localization, as a matter of fact, can be thought of as a symmetric hostage.

When the economic rationale behind technological communication is made clear and the pay-off of interacting agents, well aware of both the positive and negative effects of knowledge leakages, is taken into account, we see, in fact, that technological districts are also the result of an intentional decision to co-localize and participate in the flow of technological communication. Especially for large firms, co-localization within technological districts and active participation in local communication systems are the result of an intentional strategy of technological cooperation which often complements formal technological partnership within technological clubs.

An innovation policy aimed at strengthening the communication links among innovators, reducing communication costs, and structuring communication networks in an efficient way can be instrumental in accelerating rates of learning to communicate and in increasing the size of technological districts. Both are likely to have important positive effects in terms of the amount of collective technological knowledge which a system can generate and benefit from. The evidence suggests that markets face major problems in evolving swiftly towards an efficient design and governance mechanism of communication networks: the experience in telecommunications networks suggests that regulation is required in order to establish open interconnection rules. Knowledge-intensive service firms, however, can emerge in this context as specialized 'switches' of communication flows among firms, providing communication services to third parties and in so doing, reducing the social inefficiency of geodesic and oligarchic networks.

---

[6] Here graph theory provides important methodological help to qualify both communication networks and the role of each agent within each of them.

7

Collective Knowledge and the Dynamics of Technological Clusters: The Case of Communication Technologies

Considerable empirical evidence and important theoretical analyses in the economics of innovation have recently provided new support for the Schumpeterian notion of gales of innovation (Schumpeter, 1934: technological innovations appear in gales and cannot be analysed in isolation). A systemic approach seems necessary.

In recent years, economic analysis has paid much attention to the systemic relations among technologies. Such issues as complementarity, interrelatedness, compatibility, and interoperability among technologies have been explored and assessed. Technologies are more and more viewed as components of broader technological systems rather than self-contained, stand-alone artefacts. In this context, however, much attention has been devoted to the economic analysis of the consequences of technological systems in terms of interdependence of consumers and intertemporal choices of adoption (David, 1987). Little research, by contrast, has been done into the determinants of the sequential clustering of complementary technologies. This chapter is devoted to understanding the system dynamics of technological clusters as it emerges in the case of communication technologies.

The emergence of technological clusters is strongly associated with regional concentration of innovation activities. The important role of local knowledge externalities and hence of technological communication among researchers active in complementary research ventures which take advantage of a collective knowledge 'common' helps us to understand the growing relevance of agglomeration economies and proximity at large in the generation of new technological clusters. Proximity makes face-to-face communication and repeated interactions easier, helps build common trust in relationships, and favours the practice of symmetric gifts of information among researchers. For these reasons, clustered technological innovations in such technological systems as new communication technologies and biotechnologies are largely co-localized in a few locations worldwide, such as the Route 128 around Boston, the Silicon Valley, and the area around Cambridge in the United Kingdom.

New technological clusters may be considered the result of a complex system of radical and complementary innovations, introduced sequentially, which consist of a variety of applications of new general-purpose technologies. The direction of technological change seems, in fact, to be characterized by a sequential array of technological innovations, each of which exhibits high levels of complementarity and interrelatedness that represent at the same time product, process, and organizational innovations, as well as innovations that change the production mix of firms and their markets. This sequential array of technological innovations is itself characterized by the sequential complementarity of the innovative efforts of a variety of players, each embedded in its own localized field of activity.

New technological clusters are being implemented through a variety of technological innovations that are sequentially generated in a wide range of industries. The boundaries of such industries are more and more blurred and the flows of entry and exit of firms are fuelled both by the birth of new firms and mainly by the cross-entry and cross-exit of incumbents in related industries which try and take advantage, in terms of complementarities, of the system dynamics of technological clusters. The co-evolution of industrial dynamics and technological convergence is likely to play a key role in the eventual structure of the new technological cluster.

The emergence of such technological clusters appears to be the outcome of the sequential introduction of interdependent innovations along definable paths shaped by a number of factors: the identification of a pool of collective knowledge; technological complementarities and hence technological externalities; the advantages of learning to learn in specific areas of application; the competence accumulated by means of the dynamics of learning to do and learning to use; and, last but not least, the irreversibilities and indivisibilities associated with significant portions of the material and immaterial capital structure of existing firms at each point in time.

Dynamics of Technological Clusters 113

This chapter tries to provide a review of the new analytical evidence about technological clusters and their determinants and effects. Section 1 provides a definition of technological clusters and identifies the determinants of and constraints on the direction of technological changes, drawing upon the evidence of communication technologies. Section 2 elaborates formally the system dynamics of technological clusters. In this chapter, some effects, both at the macroeconomic and microeconomic levels, of the dynamics of technological clusters are also examined. Section 3 provides an account of the role of industrial dynamics and technological convergence in the case of new communication technologies. In Section 4, the main results of the analysis are summarized and put in a perspective which considers the scope for industrial policy interventions.

1. THE KEY FACTORS IN THE SYSTEM DYNAMICS OF TECHNOLOGICAL CLUSTERS

Technological clusters are characterized by weak separability among an array of different yet complementary technologies, many of which are currently being introduced. Such technologies are characterized by relevant complementarities such that the efficiency and profitability of introduction of each of them is strongly dependent upon the context and specifically upon the number of interrelated products and services.

A technological cluster is a set of technologies characterized by high levels of complementarities, interdependence, and interrelatedness which affect the performance and profitability of introduction and adoption of each single innovation. Technological clusters are the result of technological convergence, that is, a flow of complementary technological innovations which are introduced sequentially by a myriad of innovators and draw from a renewable pool of collective knowledge (Amendola and Gaffard, 1988; Perrin, 1991; Carlsson, 1995, 1998; Rosenberg, 1976, 1982; Helpman, 1998).

Within a technological cluster, each technology is linked to others by high levels of complementarity in terms of compatibility, interdependence, and actual interoperability. Within a technological cluster, the profitability of each technology being generated and eventually introduced in the market-place is affected by levels of compatibility and interoperability with other existing technologies and other technologies being introduced. In a technological cluster, each technology is characterized by weak indivisibilities and separabilities that consist in direct and indirect linkages among products and processes. Such linkages are relevant both on the demand and the supply side and can be defined in terms of technological complementarities which take the form of economies of scope when internalized within the boundaries of each firm and externalities when their effects are not contained within the borders of a single firm. In both cases, a variety of complementarities can be identified: knowledge complementarities, technological complementarities, pecuniary complementarities, and direct and indirect (both supply and demand) network complementarities (Hughes, 1984; David, 1987, 1992; Helpman, 1998).

Collective knowledge consists of the positive and negative externality effects on the productivity of resources invested in research and development activities in the introduction of innovation i of a number of related and complementary innovation activities conducted within the system at each point in time and in the past. Technological complementarities consist in the direct effect on the general level of total factor productivity of technology i of a number of related and complementary technologies available at each point in time. Pecuniary complementarities consist in the effects on the costs of intermediary production factors of the number of complementary activities which use the same intermediary inputs. Direct network complementarities consist, as is well known, in the effects on the utility and production function of the installed base of users of a given technology. Indirect network complementarities instead focus on the effects, both on demand and supply, of the installed base of a given technological innovation and hence of a new product or a new process on other products and processes (Antonelli, 1995, 1999*b*).

Such bundles of complementarities are relevant both *synchronically* and *diachronically*. Synchronic complementarities take place among technological innovations being introduced. Clustered innovations benefit each other when their complementarity enhances overall levels of utility and productivity. Diachronic complementarities consist in the effects of existing stocks of fixed capital on the patterns of consumption and on the profitability of introducing and adopting each technological innovation (Dixit, 1992). On the supply side, the compatibility and interoperability of a new technology with the existing stock of fixed, tangible and intangible, capital goods and skills and competencies of employees and organizations enhances its profitability because it reduces the need for the anticipated scrapping and reskilling and retraining of the human capital and organization of firms. On the demand side, the compatibility and interoperability of a new technology with existing technologies make its adoption easier, in that this reduces the switching costs for consumers and users and has positive effects on both the slope and the position of the demand curve for that product. Hence complementarities can be found both across products and processes: the former when products are jointly used, the latter when different products are manufactured using the same process. Finally, complementarities can be fully internalized so as to lead to economies of scope. Complementarities, however, can also occur between products and processes operated by different firms. In this case, they lead to externalities (David, 1975, 1985).

The emergence of technological clusters, as characterized by the specific distribution of complementary technologies, each of which is linked to the others by compatibilities, interrelatedness, interoperability, and interdependences, appears to be the outcome of the direction of the local search for new technologies. For given levels of irreversibility and sunk costs, economic systems and firms that are exposed to high levels of economic entropy are likely to react with larger levels of innovative efforts. In turn, with given levels of innovative efforts, the rate of technological changes actually introduced is influenced by the number of technological opportunities available at each point in time. Technological opportunities are strongly influenced by regional localization. Localization in regions that offer high levels of technological communication among innovative agents is a major determinant of technological opportunities.

The basic elements of the system dynamics of technological change can now be articulated. Diachronic and synchronic complementarities among technologies are, at the same time, a constraint and an opportunity for each firm, when technological change is endogenous. Complementarity is a constraint because the introduction of a new technology by a firm has direct and indirect effects on all the other firms. Their product and factor markets are affected as well as the collective knowledge base: former externalities no longer hold and new ones emerge. Prices reflect such changes only to a limited extent. Firms, exposed to the change, can either adjust to the new market conditions or react creatively and introduce new technologies in turn. Here complementarity acts as an opportunity. The larger the knowledge and network externalities and the better the communication system, the higher is the chance that each firm can introduce successful innovations and affect the market conditions of other firms in turn. The conditions for an endogenous process of technological and economic change are now set. The rate and direction of technological change seem to depend to a large extent on the capability of firms to react successfully to the challenges of the system. The distribution of superfixed production factors, and the communication and access conditions to collective knowledge, play a key role in this context.

Specifically, evidence about the conditions into which innovations are being introduced in communication technologies underlines the key role of:

(1) *Endogenous economies of scope* New communication technologies provide the most striking evidence about the pervasive role of economies of scope. Average costs are strongly influenced by the variety of products that can be produced and delivered with a common infrastructure, both tangible and intangible. This is to a large extent the result of the intentional generation and introduction of new technologies that value existing capital stocks.

(2) *Endogenous technological externalities* The local search for new technologies that make use of pre-existing and irreversible production factors lead to endogenous technological externalities when such irreversible production factors are external to firms but internal to both clusters and districts. New communication technologies provide striking evidence about the pervasive role of such endogenous technological externalities and weak divisibilities among different vintages of external production factors.

(3) *Collective knowledge* New communication technologies provide impressive evidence about the horizontal complementarity between the multi-purpose role of generic technological knowledge and the myriads of specific and idiosyncratic contexts of application.

(4) *Technological cooperation* The flourishing technological cooperation among firms, often active in diverse and yet interrelated industries, reveals the interest of firms in building together a

common base of codified generic technological knowledge which can eventually be used in a variety of non-rival uses, generating technological innovations in diverse and yet interrelated market niches. Technological cooperation takes place by means of the creation of technological clubs and co-localization within technological districts.

(5) *User–producer interactions* Accentuated user–producer interactions take place in many information and communication technologies with users increasingly interested in entering the innovation process in communication technologies because of their growing role in the production process as well as in its management. An array of important innovations is being introduced by large users, both in manufacturing and service industries, as a result of their direct impact on their current business. The interaction between producers of communication services and products and users is becoming a major source of technological innovation.

(6) *Standards* Interoperability is often the result of the standardization of products and processes. In turn, standardization can be the result of market forces, so that products are selected according to their actual levels of interoperability with other key technologies. Standardization, however, can also be the result of committees. The latter lead to *de-jure* standards while the former to *de-facto* standards. The choice of standards and the institutional context in which this takes place can have important consequences on the industrial dynamics (David and Steinmueller, 1994).

(7) *Gateway technologies* The introduction of new technologies specifically aimed at increasing interoperability and hence achieving complementarity, both among pre-existing products and processes and new ones and between new technologies being introduced, plays a key role in the system dynamics of technological clusters and can help increase their scope of application.

(8) *Usage innovations* Intentional diachronic technological externalities, that is, the effects of the purposeful complementarity between new technologies and portions of the existing infrastructure, are especially relevant in the case of production activities shaped by significant irreversibility. A significant amount of innovative activity is oriented towards the introduction of technologies that make it possible to take advantage of existing capital stocks, both tangible and intangible.

(9) *Technological rivalry* New technologies also provide clear evidence about the strong localized technological rivalry among innovations being introduced. Here, direct and indirect network externalities are very effective. The profitability of the introduction and adoption of each given technology is, in fact, strongly influenced by the context of introduction and adoption, that is, the number of complementary innovations with which the new technology can be interoperated and the size of the installed base of similar products. The selection of technologies appears to be strongly influenced by the timing of introduction as well as by the sheer levels of efficiency (Arthur, 1989).

(10) *Demand network externalities* Communication supplies a strong and self-evident example of the key role of direct and indirect network externalities on the demand side. As far as the former is concerned, it is clear in fact that the utility of the usage of each communication device is largely influenced by the variety and number of interoperable communications systems.

(11) *Supply network externalities* These are important where the efficiency of each single piece of equipment is strongly influenced by the number of other complementary pieces that can be interrelated and with which interoperability can be achieved.

(12) *Irreversibility and innovation* Production systems are often characterized by the long durability of large chunks of equipment and their increasing scope of usage. The need to keep in production significant amounts of installed fixed capital becomes an important focusing device to introducing new technologies that make it possible to create new uses for existing capital goods.

(13) *Economies of density* The important role of superfixed capital goods finds direct application in terms of economies of density, that is, the reduction of unit costs associated with the repeated use of such capital goods over time.

(14) *Incremental costs* For a given infrastructure, the cost of implementation is lower when portions of existing capital goods can be expanded so as to take advantage of additional economies of density.

(15) *Diversification and vertical integration of incumbents* Incumbents try to take advantage of the complex array of externalities which constitute the basic fabric of technological clusters so as to increase the variety of their products and hence internalize the positive effects of externalities with an active strategy of cross-entry in

related product and geographic markets (Langlois and Robertson, 1995).

(16) *Specialized entry* Innovators, on the other hand, try to make the best use of single innovations which provide some competitive advantage in a specific market niche. Eventually such niches can be implemented and become the vector of complementary innovations.

2. THE SYSTEM DYNAMICS OF TECHNOLOGICAL CLUSTERS

The rate and direction of technological change can be viewed, in the long term, as the endogenous outcome of the innovative sequential reaction of firms induced by the interplay between the irreversibility of their capital stock and economic entropy. Irreversible capital stock can be thought of as constituted by both fixed physical capital and competence and technological knowledge in well-defined and circumscribed technical fields. Collective knowledge and complementarity among technologies being introduced become relevant here to fully assess the system dynamics of technological clusters. Technological opportunities can now be specified as a system of complementarity among new technologies as well as between new technologies and existing ones which is at the origin of relevant externalities and economies of scope.

Formally, we see that from a supply-side viewpoint, a specific technology belongs to a technological system when, because of technological network complementarities, total factor productivity is functionally dependent on the levels of implementation of a number of other technologies:[1]

$$Y = A(K, L) \qquad (7.1)$$

$$A = f(N, T) \qquad (7.2)$$

where K stands, as usual, for capital, L for labour, and T for technology. N stands for the number of other complementary technologies both in place or being introduced and for the size of their installed base, and $f' > 0$, but $f'' < 0$.

---

[1] Where equation (7.1) is the specification of the general version: $Y = Y(A, K, L, a, b)$.

Pecuniary negative externalities affect the market price (p) of intermediary knowledge inputs. The unit costs of research, communication, and learning activities (R&C&L) are positively affected by the number of other technologies that make use of the same limited amount of available knowledge inputs. Formally, we see that for

$$TC = rK + wL + pR\&C\&L \tag{7.3}$$

$$p = g(N) \quad \text{with } g' > 0. \tag{7.4}$$

From a demand-side viewpoint a specific technology belongs to a technological system when in the utility function and in the related demand function, the utility of the good Y is directly affected by the number of other complementary technologies being implemented and adopted and their installed base:

$$U = (Y^a, X^b) \tag{7.5}$$

$$a = h(N) \tag{7.6}$$

where N stands for other complementary technologies both in place or being introduced and for the size of their installed base, and $dU/dN > 0$, but $dU^2/d^2N < 0$. The effect of complementary goods on the utility of each can be assumed to be positive but with a decreasing impact.

When we translate the specification of technological interdependence in terms of the profit function, from the demand side we see that the revenue equation is functionally dependent on the levels of implementation and adoption of other technologies. The demand for a good which is complementary and interdependent with others rotates rightward, with an increase in the revenue elasticity and a decrease in the price elasticity, with an effect that is stronger, the larger the impact of direct and indirect network elasticities. The same effect takes place on the supply side in terms of a reduction in costs for the twin effects of technological and pecuniary externalities.

The next step consists in the specification of the market form for complementary innovations. Following a long-standing tradition of analysis, a context of monopolistic competition with barriers to entry seems most appropriate to model Schumpeterian rivalry. Each

Dynamics of Technological Clusters 121

firm can be considered a local monopolist in a demand niche. The size of each demand niche is positively influenced by the number of complementary products.

Two categories of players must be taken into account: incumbents and new entrants. In both cases, the technological cluster is progressively enriched and articulated. Diversification of incumbents based upon latent economies of scope and entry of newcomers, able to take advantage of externalities, are the two driving processes.

Incumbents in each demand niche have a strong incentive to internalize the positive effects of latent demand, knowledge, technological, pecuniary, and production complementarities. Complementarities are the source of economies of scope which, in turn, open the way to cross-entry in adjacent markets. Endogenous economies of scope are the result of the internalization of potential complementarities of new products and production processes, both with existing ones and with other new products and processes. Extra profits earned in the original market niche are partly used to fund localized research and learning activities directed towards the identification and valorization of potential complementarities, so as to introduce new products and processes that can take advantage of existing and irreversible production factors. Incumbents try and extend their competitive advantage from one market to another by means of bundling strategies. Here the dynamics of endogenous economies of scope plays an important role and the ultimate result is growing diversification and integration, respectively in lateral and vertical markets. Organizative diseconomies of scale and scope reduce the ability of incumbents to take advantage of latent complementarities. Beyond some critical size, the unit costs of internal coordination are likely to increase rapidly, both with the sheer levels of output and specifically with the diversity of products which are being manufactured and marketed.

Latent complementarities, especially on the demand side, can be identified by newcomers. The entry of imitators in that specific market niche is impeded by barriers to entry, but the entry of product innovators in adjacent markets is free. As much evidence suggests, the rates of entry of new firms are positively correlated to the rates of introduction of innovations. The entry of new innovative firms introducing new products has the twin effects of reducing the size of the market niche for the incumbent but also of increasing it via the positive effect on the utility function for those specific products engendered by the availability of a new complementary product in the market-place.

Innovative entry plays a crucial role in this model. As a matter of fact, both cross-entry of incumbents and entry of newcomers is associated with innovation in complementary technologies. In turn, levels and rates of innovative entry depend upon access to and communication conditions within the collective pools of technological knowledge. As such, innovative entry has a strong stochastic character: it may take place or never actually be consolidated. The timing of entry is crucial. According to actual rates of entry, the full advantages of a technological system may be fully and quickly exploited, just as they may remain potential ones and the system can emerge very slowly.

In both cases, whether latent complementarities are exploited by incumbents or newcomers, it is clear that in the long term, specific research, development, and learning unit costs do increase with the number of complementary technologies being introduced.

Hence we can stylize directly the long-term effects of technological convergence on the profitability of each local monopolist so that we can define a revenue function where the positive effects on the revenue of each firm j of the complementary technology i (IRij) are, *ceteris paribus*, directly influenced by the direction of other firms' innovative efforts, in terms of the number of complementary innovations:

$$IR_{ij} = i(N) \qquad (7.7)$$

where N stands for other technologies both in place or being introduced and the size of their installed base and $dIR/dN > 0$, but $dIR^2/d^2N < 0$.

The incentive to generate, introduce, and adopt new technologies that are complementary and interdependent with others is now clear and consists in higher revenues that are functionally dependent upon the number of complementary technologies.

The specification of the number and characteristics of the technologies that constitute a technological system now becomes central in this context. According to a large empirical literature, it seems clear that technological clusters have borders and maximum density and consist of a maximum number of complementary and interrelated technologies.

The role of positive and negative knowledge complementarities is very important. Firms conducting research can benefit from the spillover of positive knowledge externalities and technological externalities stemming from the complementarity of innovative efforts of other research programmes within firms and most important, from the research programmes of other firms and other related research institutions. However, it seems also clear that the larger the number of such technologies being implemented, the lower will be the scientific opportunities to introduce further innovations.[2] Negative knowledge externalities arise in terms of declining scientific externalities due to 'overfishing in the common', that is, within the borders of the 'pool' of complementary knowledge which makes possible the generation of new technologies.

In the long term, total research costs in a technological cluster increase more than proportionately with the number of complementary technologies being introduced. Unit costs of the dedicated and specific knowledge resources that are necessary to conduct research activities within a given technological cluster increase: negative pecuniary externalities apply.

The basic elements for understanding the long-term system dynamics of technological clusters are now in place. In this context, in fact, the basic tool of the traditional (Marshallian) entry process driven by profit maximization can be applied so as to understand the direction of innovative efforts. Technological convergence, that is, the flow of introduction of complementary innovations now finds a well-defined context of understanding. The introduction of complementary innovations will take place as long as the maximum number of complementary innovations is achieved.

In this approach, technological convergence can be understood as the result of the rational behaviour of firms, both incumbents and newcomers, which decide the direction of innovative efforts. Firms

---

[2] A distinction is made here between technological externalities which stem within technological systems from the interaction of learning and innovative agents and scientific opportunities which are generated from breakthroughs in scientific knowledge. Scientific knowledge does retain elements of exogeneity.

have already decided a given level of resources to fund the introduction of innovations and to try and assess rationally the direction of their innovative efforts in terms of the typology of new products and new processes and their separability with respect to both other innovations and to the technologies already in the market-place.

Firms can decide whether to introduce innovations that are complementary to others and in which specific subset of complementarity, that is, in which technological cluster. Within this context, technological convergence can be claimed to be the endogenous result of the intentional activity of firms subject to the actual effects of the complementarities and externalities at work. The two extremes are clear: firms can decide to introduce stand-alone technologies or technologies that are complementary to a large array of other products and processes. In the latter case, the firm would contribute a technological innovation to a technological cluster that has not yet reached the maximum number of complementary technologies.

Firms become aware of the opportunities offered by complementarities in selecting the innovations they can introduce in the market-place. New products and new processes that are complementary to others offer the clear advantage of benefiting from steeper demand curves with larger revenue elasticities and larger scope for increasing total factor productivity.[3] Such an understanding, however, requires time in order to be implemented and put into practice.

At the aggregate level, the profit equation makes it possible to assess the long-term 'correct' direction of technological innovations of each firm and hence the size and density of each technological system. The correct number of firms is identified by the level at which profits are driven to zero or to average levels (AP). The long-term revenue equation can be modelled as functional in N, the number of complementary innovations and the same is true for the

---

[3] At the disaggregate level, in the short term, each firm will maximize profit for given levels of complementarity. Hence each firm j will direct the resources available into the technological system i, *ceteris paribus* the other inputs, so as to maximize the following profit function:

$$IP_{ji} = IR_{tji} - R\&C\&L_{ji} \qquad (7.8)$$

where

$$\max P \text{ for } dIR_{ji}/dQ - dR\&C\&L_{ji}/dQ = 0. \qquad (7.9)$$

cost equation for the direction of research activities. Formally, we can write:

$$\max N \text{ for } IR_i(N) - R\&C\&L_i(N) = AP \qquad (7.10)$$

These elements should now be sufficient to provide a general view of the system of incentives and opportunities that the emergence of technological clusters provides for the conduct of firms. Firms have, in fact, a clear incentive to concentrate their innovative efforts within a technological cluster as long as its maximum size is not attained.

One further important step can be made. Let us make explicit the role of time t in the relationship between the number of complementary technologies and their related installed base and the total factor productivity level. The introduction of the complementary technological innovations which constitute the new technological cluster emerges gradually over time and affects the efficiency of the system. Hence we can specify the historic dimension of the evolution of profits as follows:

$$IP_{ti} = m(N_{ti}) \qquad (7.11)$$

Specifically, we see that, because of the positive slope of research costs associated with the number of complementary innovations and the positive, but decreasing slope of the revenues associated with the number of complementary innovations, for each technological system i, the profit (IP_{ti}) stemming from the direction of technological change towards the introduction of complementary technologies within the technology system i can be shaped as a quadratic function in N_t:

$$IP_{ti} = aN_{ti} - bN_{ti}^2 \qquad (7.12)$$

Let us now assume that at each point in time the resources devoted to research, development, and learning activities, finalized to introduce complementary innovations within the technological system i ($R\&C\&L_i$), are a function (v) of the net revenue stemming from the introduction of complementary innovations.

This is the result of both the conduct of incumbents who learn about the opportunities of the technological cluster i, are able to take advantage of latent economies of scope with the implementation of strategies of cross-entry and bundling, and the entry by new firms attracted by the profits of complementarity which are available in the form of externalities. Hence:

$$R\&C\&L_{ti} = v(IP_{ti}) \tag{7.13}$$

In turn, the introduction of each additional complementary innovation (n) is a function of the amount of R&C&L expenses directed towards the system i:

$$n_{ti} = w(R\&C\&L_{ti}) \tag{7.14}$$

The substitution of equations (7.13) and (7.14) into equation (7.15) leads to the following:

$$n_{ti} = p(v(aN_{ti} - bN_{ti}^2)) \tag{7.15}$$

Because $n = dN/dt$, we find the traditional relationship of a rate of growth which depends upon the quadratic specification of the stock which has its solution in the standard logistic function.

It is now clear that the more rapid is entry, the shorter is the completion of the technological system. Actually if and when entry does not take place, consolidation of the technological system never takes place and complementarities remain latent.

The interpretative framework implemented so far is consistent with the empirical evidence. Extensive empirical evidence suggests, in fact, that the time profile of the completion of a technological cluster can easily be approximated by a logistic distribution which exhibits a long phase of slow progress, a period of rapid introduction of new complementary technologies and actual emergence of the technological system, and eventually an extended period of approximation to the asymptotic number of complementary technologies.

The approach elaborated so far has important macroeconomic implications. It is clear, in fact, that, because total factor productivity levels, pecuniary externalities, and consumer's surplus are all influenced by the size of the technological cluster, the larger is N, the number of complementary technologies, the larger are the general levels of efficiency of the economic system. From a dynamic viewpoint, it is also clear that such general efficiency grows, along a logistic time path, as long as the maximum size of the technological system is achieved and the longer is the time spell during which the dynamics of technological convergence lasts (Arthur, 1994).

The consequences of the logistic path along which technological clusters emerge is also relevant in this context. Total factor productivity growth will actually take place along such an S-shaped time profile, with evident effects in terms of the time distribution of the rates of growth of the system at large.

In this context, access conditions to renewable pools of collective knowledge and technological communication play a key role. The larger their positive effects in terms of technological opportunities and the size of the renewable common pool of accessible knowledge inputs, the larger is the size of the technological system. In turn, the smaller are the negative effects of declining technological opportunities, the larger is the size of the technological system with all the positive effects already considered (Carlsson and Stankiewitz, 1991).

3. INDUSTRIAL DYNAMICS AND NEW TECHNOLOGICAL CLUSTERS: THE EVIDENCE OF NEW COMMUNICATION TECHNOLOGIES

New communication technologies are today at the core of the direction of technological change in advanced countries. They are being implemented through a variety of technological innovations that are sequentially generated in a wide variety of industries such as telecommunications, television, software, electronics, microprocessors, space industries. The boundaries of such industries are more and more blurred by the flows of entry and exit of firms.

Industrial dynamics is fuelled both by the birth of new firms and mainly by the cross-entry and cross-exit of incumbents in related industries. Newcomers try and take advantage of complementarities in the form of externalities—mainly demand externalities when they are able to spot latent complementarities among products. Incumbents try and take advantage of the system dynamics of technological clusters provided by complementarities in terms of endogenous economies of scope.

At the same time, the composition of each of the industries as such is being challenged by a variety of technological innovations. The co-evolution of industrial dynamics and technological convergence is likely to play a key role in the eventual structure of the new communication cluster at large (Nelson, 1998; Amendola et al., 1999).

In this context, telecommunications play a key role on many counts. First, telecommunications are the main carrier of a variety of communication and information products that enter the production process of a wider and wider array of other industries and economic activities at large. The transition to the new knowledge economy seems to rely more and more on the quality and variety of advanced telecommunications services. Relatedly, the sharp reduction in the price for telecommunications services, both in nominal and especially in hedonic terms, because of the impressive increase in transmission capacity in terms of data and images, has powerful effects on the rest of the technological system in that it makes available at low costs remote resources which can be easily used at distance. Low communication costs favour the division of labour and competition among products and processes, reducing the role of distance for a wide variety of products and especially services. Accessing, storing, retrieving, processing information is now made possible by computer communication at costs that are so low as to engender significant effects in terms of transaction costs, coordination costs within large organizations, and location costs (Antonelli, 1999b).

In telecommunications, traditional articulation in transmission, distribution, switching, and signalling is being challenged by the introduction of mobile telephony which uses new transmission capabilities, at least for part of the traffic. The diffusion of data communication among companies and in the household, because of the Internet, stresses the role of fixed distribution and transmission infrastructure and the need to enlarge transmission capacity to carry high-speed data as well as images. The traditional boundaries between communication and broadcasting are being challenged by the new capacity to carry images on broadband fibre optics, as well as by the new opportunities of interactive television. The same is true for informatics, publishing, and financial products where the division of labour between contents and means of transportation is more and more blurred due to the idiosyncratic specification of the communications software. But at the same time, the new generations of mobile telephony and space technology can provide new alternative routes for transmission and distribution with low orbit satellites which perform at the same time switching, signalling, and

Dynamics of Technological Clusters 129

transmission functions for voice, data, and images. The diffusion of personal computers relies more and more on access to the Internet and related services which make possible the introduction of low-cost computers. These latter can, in fact, rely upon complementary processing capability located at a distance but accessible via cables.

Specifically, competition and technological change in telecommunications markets appear to be key to understanding the role of industrial dynamics in assessing the emergence of the new communication system. The increasing variety of communication products on the one hand and their strong complementarity both on the supply and demand side has a number of important and yet contradictory effects.

Telecommunications have been facing the introduction of drastic technological changes since the late 1960s after the advent of microelectronics and the related flow of product and process innovations. Microelectronics has, in fact, changed the conditions for the production and use of telecommunications drastically. The rapid introduction of digital technologies has changed the production process completely not only with huge reductions in costs, but also and mainly with a drastic fall in minimum thresholds for efficient networks. Microelectronics and its applications to computers and informatics, however, have also changed dramatically the conditions for usage of telecommunications systems, introducing data and eventually image communication.

The merging of computers, telecommunications, and television can itself be considered the cause and the consequence of a radical technological change in telecommunications. The advent of computer communication especially can be seen as the major factor driving a rapid and continual shift in demand for telecommunications services. The rapid increase in demand for telecommunications services has forced firms to elaborate new growth strategies and to actively search for new technologies within a limited technological range of options which is as close as possible to the original endowment of irreversible, superfixed production factors. Differentiated technological strategies have progressively emerged, clustering, respectively, around the fixed network endowment of incumbents and the radio-communication opportunities.

As a result, technological change has been especially rapid and radical in telecommunications in the last thirty years, with investment in the full range of technical functions of telecommunications: transmission, switching, distribution, signalling, and network management. Each major technological innovation introduced in telecommunications in the last thirty years has changed the shape of the network and the complementarity among the different components of the network. Technological change has affected in depth not only the market position of each operator in the network but also the technical and organizational features of the architecture of the network. Each of these innovations deserves to be analysed as the outcome of the localized effort of well-defined groups of actors, changing their technology and consequently their role within the network, their markets, and their profitability.

Analysis of the most important technological innovations introduced in the telecommunications industry in the last thirty years enables identification of two clusters: (1) centripetal technologies that enhance the relevance of economies of scale, scope, and density, stemming from the existing superfixed capital infrastructure as constituted by the existing network, mainly introduced by network operators in order to enhance the dynamic efficiency of their actual production mix, and more specifically of the superfixed production factors such as the existing network, articulated in cables, switching, and transmission equipment and (2) centrifugal technologies that reduce the relevance of technical economies of scale and density; specializing technologies that reduce the role of interfunctional telecommunications economies of scope, and segmental technologies that reduce the role of network externalities, all mainly introduced by large users and new mobile competitors which rely on wireless communication as a way to reduce the strong competitive advantage of incumbent network operators.

The mapping of major technological changes introduced in telecommunications in the last forty years makes it possible to attempt to understand the direction of technological change in this area as a result of the contrasting innovative efforts of a variety of players characterized by their localized search for new technologies.

The cluster of technological changes introduced in the 1950s and 1960s reflected the market conditions of the telephone industry from the 1920s until the late 1960s, that is, the 'perfect' natural monopoly, and can be thought to have been influenced by the 'centripetal' efforts of the natural monopolist to reproduce and extend the conditions of the monopoly by the introduction of innovations

Dynamics of Technological Clusters 131

that reinforced the dynamics of increasing returns. This technological path was clearly shaped by the effort to increase the intensity of use of the existing superfixed capital infrastructure centred upon the existing network. The focus of the research strategy was clearly directed towards the switching and signalling functions.

In the early 1970s, along with the introduction of other new communication technologies, the direction of technological change in telecommunication services entered a new phase, fuelled by the emerging relevance of new dedicated services for large users and new market opportunities in mobile telephony. This new phase made the institutional set-up and the established organization of the industry progressively out of date. The new cluster of technological changes introduced in the telecommunication service industry from the late 1960s until the mid-1980s had, rather, all the characteristics of a localized process of innovation led by large advanced users and new mobile competitors. The focus of technological change was clearly directed towards new transmission systems based upon radio links aligned in the effort to minimize the requirements of the Hertzian space and to reduce the limitations of the bandwidth.

Space technology is also an important area of activity in this context for the possibilities it offers to better direct the ever-increasing flow of mobile traffic. Space technology finds in this context its first, strong, and strategic area of civil application.

Data communication has played a key role in this context. Initially, data communication concerned limited groups of very large users that had low levels of network externalities. These were very sensitive to the high levels of long-distance tariffs for their large quantities of communication. The focusing device here was on the intrinsic opportunities for cream-skimming that are almost necessarily built into markets characterized by network externalities and price discrimination. The incentives to price-skimming entry became all the stronger, in a context characterized by price-discrimination schemes, based mainly on distance, after the introduction of new transmission technologies and the diffusion of dedicated lines. The segmentation of overall demand into niches and the identification of classes of customers with low levels of demand externalities *per se*, without any effects on the cost side, offered important market opportunities. New communication media, such as dedicated data-communication systems, actually made it possible to better identify the needs of some important and growing groups of customers.

The pressure of large customers, stirred by the high levels of tariffs paid to telecommunication carriers for their increasing levels of data communication and, more important, the opportunity to reap the important quasi-rents associated with the innovative usage of telecommunication services, may be considered one of the strongest factors influencing both the enhanced rate of introduction of technological innovations since the late 1960s, and the shift in the direction of technological change following the introduction of centrifugal innovations. This process eventually led, at least in the USA, not only to divestiture but, more important, to segmentation and specialization of the centralized network into a web of special-purpose networks. Network externalities, especially on the demand side, exert a much smaller appeal for dedicated and specialized data networks where communication is limited to small groups of advanced and sophisticated users, consisting mainly of large computers located on the premises of large companies. The fast diffusion of Intranets within large corporations has been a major centrifugal force driving towards the creation of a variety of dedicated networks which did not need to be interconnected. The large endowments of electronic hardware of large corporations and their acquired competence in managing application software played a major role here.

Data communication had first been introduced by large multinational companies and eventually diffused to large corporations to coordinate their global and multi-site operations by means of Intranets characterized by a limited number of connected users. Idiosyncratic standards, tailored for internal communication and computing needs, also played a role: traditional telecommunications could not easily match the huge variety of dedicated standards of each user. The architecture and management of such Intranets were provided internally: telecommunications carriers were asked to provide as little as transmission. The high levels of telecommunications tariffs, especially on international routes, forced multinational companies to try and reduce to a minimum the costs of their Intranets through the use of dedicated lines. Irreversibility here acted as a basic sorting device to generate new economies of scope: the existing stock of competence in software, hardware, and in managing complex databases was readily extended to new and innovative communication functions.

Divestiture in the USA and massive privatization and liberalization in Europe provided the institutional context which enabled the entry of new specialized competitors and diversification of large corporations in telecommunications services.

The centrifugal effects of technological change and industrial dynamics are also relevant in the 1980s with respect to the traditionally strong vertical ties between carriers and hardware manufacturers. In the 1980s, radical vertical disintegration between telecommunications services providers and telecommunications hardware manufacturers in fact took place. The vertical links between manufacturers and carriers were put under strain because of the increasing variety of new customers, including specialized newcomers and large users. At the same time, on the supply side, new technologies were often supplied by newcomers in the hardware industries and a wave of mergers and acquisitions in the hardware industries severed traditional user–producer relations.

Gradually, however, during the 1980s, Intranets grew into Extranets. Large corporations took full advantage of the scope for application of their own internal data communication networks to manage distribution systems and relations with their suppliers. Intranets slowly became Extranets. Huge business-to-business data communication flows were growing in order to make possible the integrated logistics required by just-in-time management techniques. Suppliers were required to provide timely intermediary inputs. Deliveries were planned and inspected via data communication. Advanced subcontracting was to an increasing extent implemented with data communication systems based upon more and more sophisticated Extranets. Co-engineering and shared research and development activities were also implemented online with a growing number of parties involved. The online management of financial services in particular quickly became an area of exponential growth. This evolution had major consequences for the architecture of telecommunication networks. Interoperability of standards became a central issue. Internal switches had to face growing traffic. Telecommunications carriers were able to take advantage of their competence in managing the ever-increasing complexity of the webs of Extranets. Specialized networks declined and traditional telecommunications carriers learnt to supply bundles of dedicated services, tailored to the needs of the Intra- and Extranets of large corporations, especially of multinationals, with a global range of operations.

The social need for an advanced universal network became evident again after years of theoretical neglect. The powerful dynamics of network externalities was again in place. Data communication, however, was now the driving force. The speed and quality of digital communication became a central issue. The capillarity of Extranets and their varying configurations, with flows of entries and exits, reduced drastically the viability of dedicated and personalized networks.

Telecommunications carriers could now try and merge the fast-growing traffic engendered by data communication and traditional voice telephony. Since the mid-1980s, technological change has been progressively shaped by such centripetal and integrating technological innovations as ISDN and BISDN, ATM. The former applications of digital technology make it easier to find centralized solutions not only to the management of administrative problems (essentially billing), but also to technical problems related to traffic management (switching, routing, and so on), and the supply of increasingly advanced services.

The interplay between centrifugal and centripetal forces in shaping the direction of technological change in telecommunications has been in place, not only in transition between voice telephony and digital communication, but also in the duel between mobile and fixed telephony. In the late 1980s, mobile telephony became the new technological axis upon which much technological change was aligned in the effort to make entry easier and to minimize the need for interconnection with the existing fixed network. Mobile telephony has been the main tool for newcomers in telecommunications to enter the market and overcome the substantial barriers to entry in traditional voice telephony. Mobile telephony was the enabling technology for the entry of competitors in newly liberalized telecommunications markets. Cross-entries in other incumbents' market were based upon mobile telephony in most European markets.

This new technology has been the object of intense incremental innovation, paving the way to putting in place a completely new technological system centred upon the systematic use of the Hertzian space as the main communication infrastructure. It is useful to recall, however, that the basic technologies had been available for more than sixty years: mobile telephony had been in use by the police, the army, and taxi drivers for many years.

This trend has important implications with respect to the organization of the innovation process within the telecommunication services industry. In recent years, technological change in the telecommunications industry, traditionally determined by the interaction of telecommunication equipment manufacturers and incumbent telecommunication network operators, is now increasingly characterized by interaction between large users, fixed equipment manufacturers, mobile equipment manufacturers, and network operators. Technological change in telecommunications and new communication technologies as a whole is more and more the result of the technological strategies of different groups of firms, each pursuing their own market strategy and building upon their localized, specific technological capability. This new trend has drastically altered the traditional centripetal direction of innovations, adding strong centrifugal effects.

Since the late 1980s, divestiture in the USA and the wave of liberalizations and privatizations in European markets have allowed the entry of a variety of competitors, increasing technological pluralism. Cross-entries, often based on joint ventures or mergers, acquisitions, and multinational growth, are continual in this period and can be considered as attempts to try and assess which is the best combination of new technologies and new markets. Hardware companies enter telecommunications and software and vice versa (Mowery, 1996). Entertainment and television companies enter telecommunications and vice versa. Even financial service corporations try to assess the scope for new business diversification into electronic communication systems. Large manufacturing corporations internalize the provision of advanced telecommunications services and eventually become direct competitors to telecommunications carriers (Preissl, 1995). Cross-entries take place, especially in telecommunications, both from upstream industries which are interested in assessing the scope for vertical integration into services, and from downstream users which are interested in assessing the scope for upstream integration into telecommunications services.

The introduction and subsequent fast diffusion in the early 1990s of the Internet and the World Wide Web marked a major shift in the direction of technological change, in that it paved the way to mass usage of the fixed network by households and related increased interest by manufacturing and service firms in interacting online with a large base of customers. The advent of electronic commerce induces a major shift in the use of data communication from business-to-business communications flows to business-to-consumers communication flows, changing drastically the role and scope of economic interest of the existing distribution network. Network externalities on both the demand and supply sides again play a major role, as in traditional voice telephony. The introduction of enhanced broadband fibre optics and the continual reduction in their costs and market prices, engendered by economies of scale in production and increased competition in the market-place, favours the implementation of major backbones revitalizing the role of the basic network. This trend is further enhanced and fuelled by the continual introduction of new signalling and compression technologies (ADSL) which make possible relevant economies of density and scope in the usage of the existing coaxial cable network to carry data and images at high speed. Similarly, the advent of digital TV opens up the possibility of using the existing infrastructure in a new way for the distribution of entertainment and specialist interactive TV programmes.

A new duel emerges at this time between content providers and communication carriers. Like the entertainment industry, attracted by new opportunities to take advantage of existing coaxial cables specialized in TV traffic, online service providers try and establish their own networks. The new industry of Internet services providers was created through the discovery of the huge scope for electronic commerce and the high costs of telecommunications services as an intermediary input. Content providers enter the industry and are induced to create their own long-distance networks to save on high interconnection tariffs. Fibre optics technology provide an opportunity to newcomers while ADSL supplies important scope for diversification to cable-TV operators.

In the telecommunications market-place, a wave of mergers and acquisitions takes place, fed by the cross-entry of specialized carriers, respectively in fixed voice, mobile, and data. The introduction of major centripetal technologies parallels this process: Internet Protocol on the fixed network and UMTS in mobile telephony characterize this new phase of technological convergence which brings together voice, mobile, and data communication into a single network. Once again, institutional change, especially in the USA with the Telecommunications Act in 1996, paves the way for multi-media competition, favouring the creation of multi-purpose networks. The new emerging multi-purpose telecommunications networks are now the platform for the distribution of a variety of products, including financial and entertainment services. Relevant economies of scope in multi-service networks are the main tool for incumbents to build a competitive advantage compared with content providers eager to integrate vertically into bulk transmission while taking advantage of interconnection rights, made available by regulation authorities.

This dynamics of technological convergence and divergence is leading the industrial organization of telecommunications towards new tentative structures, where voice communication is more and more migrating towards mobile communications systems while fixed telephony is becoming the central infrastructure to deliver data communication, Internet services, and digital TV.

The co-evolution of industrial dynamics and localized technological change is very strong here, with a continual interplay between market and technological strategies. Incumbents try to resist the entry process by introducing centripetal technologies that build upon their superfixed tangible and intangible infrastructure. Entrants, relying on their different mix of economies of scope and density, try to build a competitive advantage by introducing centrifugal and specializing technological changes that enhance the economic value of different assets of tangible and intangible fixed capital, based mainly in electronics and service industries.

New communication products offer the opportunity for niche entry by new competitors, especially when the traditional price discrimination strategy which used distance to differentiate voice telephony has not yet been reconciled with actual costs. Here, clearly, long-distance communication products offer huge opportunities for specialized entry by competitors which can reap the large mark-ups partly used by incumbents to refund the losses engendered by traditional low-tariff local calls and universal service obligations. Cellular telephony, as well as long-distance data communication, offer similar opportunities, especially when regulatory bodies have introduced mandatory interconnection at marginal costs.

The traditional boundaries of the telecommunications industry are radically blurred and a variety of distinct industries can be identified. The entry process and the related specialization of firms can quickly lead to the identification of no less than twelve separate industries, with low levels of internalization of externalities which can be managed in terms of regulated prices and standardization processes. Regulatory agencies and standardization committees can emerge as intermediary institutions which provide the market with the necessary coordination for a variety of interdependencies which cannot be fully cleared in the market-place through the market mechanism.

In such a context, the following separate industries, defined in terms of a sufficient specification of the final product and the production process, can be considered: (1) local voice telephony; (2) long-range voice telephony; (3) multi-product transmission services providing access to the rest of the system; (4) broadcasting and television; (5) information and entertainment services; (6) bulk data communication for large users; (7) Internet service providers; (8) cellular telephony; (9) distributed software systems; (10) information services such as remote access to data banks, remote processing; (11) electronic publishing; (12) electronic finance.

In each industry, clearly, a variety of independent companies can operate with significant market segmentation in terms of quality and location. The introduction of quality as a factor of market segmentation and hence price discrimination paves the way for a variety of low-cost low-quality products which can multiply the number of regional market niches.

An alternative landscape, however, can emerge: one where all the industries listed above can easily collapse into one large industry, where only a limited number of global, diversified, and vertically integrated players can survive. In such a case, the provision of a large array of services, including entertainment, as well as distributed software, would be centralized by a limited number of large companies. The new swing in the evolution of advanced telecommunications brought about by the diffusion and implementation of the Internet and broadband optic fibres seems to favour this second alternative (Fransmann, 1994*a*, 1994*b*).

A battle of systems is nowadays taking place between two competing technological sub-clusters, each of which is being implemented and enhanced by the continual innovative efforts of different groups of firms characterized by different endowments of superfixed material and immaterial capital stocks. In parallel, accelerated industrial dynamics is in place with the continual entry of a variety of firms—often cross-entries of incumbents in related industries such as electronic hardware and software, as well as TV companies in specialized niches.

Two alternative paths are being created, one centred upon the search for complementarities between the new technologies and existing infrastructure; and a rival one, mainly based upon the search for new complementarities with the current needs of large users and mobile-service providers. While the outcome of such a battle of sub-clusters seems far from clear, one policy conclusion may be drawn from the model of localized technological change so far implemented and the specific analysis of the evolution of technological changes being introduced in telecommunications. Once more, the conditions for static optimization and dynamic efficiency may be at odds. Plurality of firms, contendibility, and entry in telecommunications are important conditions for dynamic efficiency, yet the conditions of interconnection and interoperability, both between and within networks, should be clearly defined. Entry is often spurred by the sheer divergence between interconnection costs and final prices.

Evidence from the communication technologies system suggests that the dynamics of the technological system is fuelled by the interaction between the industrial dynamics of entry by new competitors and exit, specialization, diversification, and vertical integration of incumbents, and the direction of technological convergence so as to define an architectural and structural change where the reconfiguration of existing product and process technologies and the performances of both incumbents and new competitors are strongly interdependent (Henderson and Clark, 1990).

4. CONCLUSIONS

The system dynamics of technological clusters is a fascinating area of inquiry and an important tool for understanding the results of the determinants and effects of the introduction of Schumpeterian gales of innovations. When irreversibility, weak divisibility, and low separability among technologies and high levels of complementarity matter because of the availability of pools of collective knowledge, technological innovations cannot be examined in isolation. Strong systemic effects are to be taken into account, both synchronically and diachronically.

Technological clusters are complementary to regional innovation systems and their evolution is strongly intertwined. Within technological districts, firms are likely to have high levels of technological opportunities in generating technological innovations that belong to the same technological cluster. On the other hand, incentives and growth opportunities for firms active in a single technological cluster and co-localized within a technological district specializing in complementary technological knowledge are clearly very strong.

The direction, as well as the rate, of technological change are largely endogenous to the behaviour of single firms within regional and technological systems and to the selection process built into the competitive arena. Technological changes are introduced sequentially by firms facing changes in their economic environment which require significant efforts and additional costs in terms of the search for alternative techniques, reskilling of manpower, and additional resources for purchasing flexible inputs. Switching costs can be especially relevant when the production process of firms is shaped by important irreversibilities, indivisibilities, and non-malleability of the capital stock, which in turn can be both material and immaterial.

For high levels of irreversibilities and indivisibilities which reduce and make expensive any movement along the existing technical isoquants, firms are induced to explore the scope for introducing new technologies. This localized search for new and better technologies is induced by the need to cope with the changes in the economic environment and is bounded by the competence and knowledge acquired in specific routines, products, and processes. The outcome is the sequential introduction of localized technological changes which insist upon a limited technological space gravitating around constraints imposed by the amount of switching costs.

When innovation is induced by the irreversibility of production factors which can be characterized as superfixed capital stock, new technologies that are both compatible and complementary with the technical features of the existing capital stock are likely to be introduced. In these circumstances, a dynamic process is likely to take place where, for any change in products and factors markets, firms are likely to react by means of the introduction of new technologies

Dynamics of Technological Clusters 141

that are sequentially localized and hence diachronically complementary to existing technologies. The search for complementarity is, moreover, extended synchronically, towards other technologies being introduced. The dynamics of economies of scope and externalities, both stemming from complementarities, becomes central in this context. Economies of scope cannot be considered to be a windfall or exogenous manna, but rather the actual result of the introduction of new technologies that are complementary with existing ones. The direction of technological change pursued by each firm leads to the search for and eventual introduction of new technologies that make use of existing portions of the capital stock and rely upon the competence acquired in using the existing technologies. By the same token, complementarities with other technologies are also the result of intentional decision-making by firms which try and benefit from available collective knowledge, accessed by means of technological communication channels. Innovative entry, both of incumbents and of newcomers, plays a central role in the analysis of the emergence and completion of new technological systems.

Innovative entry, however, is not obvious. It may take place or not; it may be rapid or slow. The access to and the communication conditions within the pools of collective knowledge are themselves key factors in determining rates of entry into new technological systems.

The aggregate outcome of the repeated interactions between changes in products and factors markets, irreversibility, localized and collective technological knowledge, innovative entry leads to the gradual emergence of new technological clusters characterized by important complementarities, compatibilities, and interoperabilities among different technologies.

This interpretative framework can be useful for understanding industrial dynamics as an emerging battle of technological subclusters, each of which is aligned along the direction shaped by the sequential search for complementarities and compatibility among new technologies as well as between new technologies and the existing sunk structure of the capital stock of incumbent fixed network operators and large users together with new mobile operators, respectively. In such conditions, the sequence of technological changes being introduced and of the strategic conduct of firms becomes a key element in understanding the system dynamics of the cluster and eventual outcomes (Antonelli, 1995, 1999*b*).

When a new technological cluster is being put in place, economic systems are likely to undergo a phase of rapid and drastic structural change with significant alterations of market conditions and reshaping of the borders between industries and firms. The full shape of the input–output matrix of the economic system is radically affected, with important consequences in terms of the relative prices of intermediary inputs. During such transitional phases, the basic conditions for perfect competition are clearly put under stress and the performance of market prices as efficient mechanisms for directing and assessing the division of labour and the distribution of revenue falls into question (Klepper, 1997).

Moreover, the system dynamics of technological clusters and industrial dynamics interact so as to affect both the market selection of new technologies and the performance of firms. The complementarity and interdependence between existing technologies and new technologies, as well as among the technologies being introduced at the same time, affect the market conduct of firms and the profitability of adoption and introduction of other technologies. The coevolution between industrial dynamics and technological convergence generates radical economic and technological changes which are difficult to assess and to guide. The competitive advantage of new entrants is often based upon the specific mix of technologies and products each firm is able to command and the specific origin of the economies of scope that are at play. In this context, because firms are multi-product, multi-technology, as well as multinational, it is difficult to discriminate between the selection of technologies and the selection of firms. In such a context, the failure and the success of each technology and each firm seems to be highly dependent upon the specific context and the timing of introduction and economic action. The emergence of technological clusters appears, in fact, as a dynamic process where equilibria are all but stable and the basic economic and technological attractors keep changing under the pressure of the continual but potential entry of competitors and mainly as a result of the sequential introduction of complementary innovations which redefine the profitability of each firm and each technology (Amendola and Gaffard, 1988).

This approach makes it possible to shed some light upon one more important aspect: the 'efficiency' of the market selection of new rival technologies. The market selection of new technologies is, in fact, based upon a comparison between the average total costs of firms using rival technologies: firms and technologies with higher average costs are sorted out in the competitive process. This implies that, in the competitive process, the selection of new localized technologies is heavily influenced by the pre-existing capital structure of incumbents. Specifically, we see that new technologies can appear to be less efficient than others that take advantage of a larger endowment of superfixed capital stocks already in place and are better valorized by alternative technologies. In this context, the market selection of new technologies is likely to be significantly affected by the (perverse) effects of dynamic 'lock-ins' which reproduce themselves over time because of the long-lasting effects of superfixed and often sunk production factors.

Macroeconomic rates of growth are likely to be shaped by strong non-linear effects, especially when increasing returns and the creation of monopolistic market forms are the ultimate effects of the system dynamics of technological clusters. Competitivity of regional innovation and economic systems is likely to be affected to an even larger extent, especially if international monopolistic market forms are likely to emerge and assume long-term characteristics of irreversibility and structural hysteresis.

Such long-term effects, especially in terms of the consolidation of new key sectors and key companies whose performances are likely to have profound effects on the rest of the system in terms of the relative price of intermediate products over a broad range of usage and dynamic activation effects in terms of levels of investments, rates of introduction of further technologies, and hence rates of increase of total factor productivity levels and ultimately employment and revenue levels, have already been found in the economic history of the twentieth century.

The complexity of such dynamics and their effects in terms of performance at the system level stress the need for a deeper understanding of the elements at play. It is clear, however, that economic policy needs to be directed at such dynamics and that there should be close scrutiny of the behaviour of firms in the market-place at governmental level, with the general aim of understanding not only the short-term effects of the conduct of firms but also the longer-term consequences with respect to the characteristics of the technological system as a whole.

In this context, the scope of a long-term industrial policy able to define a hierarchy of goals and to coordinate the conduct of large players emerges. A dynamic price-cap regulation which is able to design a time trajectory of reductions in tariffs and which can hence promote the timely transfer of all increases in total factor productivity, stemming from the introduction of innovations to users and consumers, is absolutely necessary.

A proactive innovation policy, aimed at enhancing the positive effects of positive knowledge externalities and reducing the negative effects of declining scientific opportunities, can have important macroeconomic results. All the reductions in the costs of research activities and related actions which favour the emergence of pools of collective technological knowledge and hence strengthen the cohesion of emerging technological clusters have, in fact, the ultimate effects of increasing the size of technological clusters and hence the absolute levels of asymptotic efficiency of the economic system as a whole. All policy interventions to reduce the height of barriers to entry and to mobility can play a key role in fostering rates of consolidation of new technology systems.

Interoperability between technologies seems an important requirement to stimulate the rate of introduction of technological change in the context of technological pluralism, one where the innovative efforts of both incumbents and newcomers can be praised, valorized, and sustained. The former are bound to innovate along the technological paths shaped by their own endowment of superfixed production factors; the latter appear more keen to appreciate the new opportunities engendered by changes in exogenous scientific opportunities, new uses, and hence new trends in demand. Contendibility should become a basic guideline for industrial policy when technological clusters are at play: technological pluralism is in fact strongly associated with conditions for entry.

Interoperability, especially when based on technological interfaces and gateways, introduced purposely, can blend different technological paths in a broader technological system.

The identification of potential technological clusters and the assessment of the scope for their development across the different segments of the European economy, can become an important area for public intervention. Public support can be provided only when and if a number of key factors, such as the quality and intensity of interdependence and externality effects, are in place and are likely to engender the dynamics of increasing returns.

The implementation of these guidelines in a context-based and selective innovation policy can become an important tool to favour the specialization of the European economy around clearly identified technological clusters so as to increase its dynamic efficiency.

8

The Dynamics of Knowledge Internalization: The Case of Fiat in the Technological District of Turin in the Mechanical Engineering Cluster

This chapter explores the dynamics of the long-term growth of a firm as shaped by the interaction between innovation, growth, and technological opportunities stemming from the dynamics of technological complementarities in a technological district and a technological cluster. The aim of this chapter is to develop the implications, in terms of a theory of the firm, of the hypothesis outlined in previous chapters about the endogenous dynamics of technological complementarities.

The object of the empirical analysis are the long-term determinants and effects of technological innovation in the historic analysis of a relevant case study in a well-defined regional and technological context. The specific object of the analysis is the growth of Fiat, the largest Italian company, specializing in automobiles, trucks, and transportation equipment in the years 1900–70. The context is the strong technological district of Turin and the emerging technological cluster centred around engineering technologies.

This seventy-year time span of empirical analysis provides a unique opportunity to test longitudinally the interactions between the pressure exerted by demand, the dynamics of economies of growth, and the introduction of localized technological changes, fed by the systematic effort to internalize relevant regional and technological knowledge externalities in shaping the growth of output and productivity of a firm.

In so doing, this case study provides strong empirical evidence on the pervasive role of technological complementarities and their effects in terms of local externalities, economies of scope, and

The Case of Fiat 147

economies of growth at a microeconomic level. The rich and detailed data set available makes it possible to test an articulated set of hypotheses which disentangle the dynamics of economies of growth and reconcile evidence about increasing returns and technological change.

The chapter is organized as follows. Section 1 recalls the microeconomic notion of economies of growth and presents an appreciative model based upon the recursive relationship between output growth and technological change, building upon the notions of irreversibility, localized knowledge, localized technological change, and technological complementarities. Section 2 provides a detailed study of the character of the process of technological knowledge generation and introduction of technological changes based upon interviews, patent data, and econometric analysis in the case of FIAT. In Section 3, the main results are considered and put in perspective.

1. THE MICROECONOMICS OF LOCALIZED TECHNOLOGICAL KNOWLEDGE AND CORPORATE GROWTH

Edith Penrose and Alfred Chandler have provided a rich, qualitative analysis of the dynamics of economies of growth at the corporate level, elaborating upon the notions of competence and organizational capabilities. The accumulation of intangible capital, consisting of experience and competence, as highlighted by Penrose (1959/1980) and Alfred Chandler (1990, 1992), and the related introduction of localized technological change (Antonelli, 1995, 1999*a*) are viewed as the key factors shaping the dynamic interaction between output growth and productivity growth.

Penrose (1959) articulates the hypothesis that firms are able to generate internal competence, defined as the capability to accumulate unused productive services which can lead to increasing levels of efficiency that are at the same time the cause of growth and its consequence. Increasing efficiency is generated by means of the accumulation of competence, based upon processes of learning by doing and learning by using, which feed the growth of output with less than proportionate levels of increase of inputs. As Edith Penrose notes:

Economies of growth are the internal economies available to an individual firm which make expansion profitable in particular directions. They are derived from the unique collection of productive services available to it, and create for that firm a differential advantage over other firms in putting on the market new products or increased quantities of old products. At any time the availability of such economies is the result of the process, discussed in the previous chapter, by which unused productive services are continually created within the firm. (Penrose, 1959/1980: 99)

The notion of economies of growth parallels in many ways the macroeconomic approach of Kaldor (1957) and provides a microeconomic rationale for understanding the relationship between rates of growth of output and total factor productivity. In this context, it seems important to understand how the growth of output actually leads to the continual 'creation of new productive services' and specifically to the introduction of innovations.

Following Alfred Chandler, this chapter elaborates the hypothesis that the growth of output forces firms to make the best use of available resources and to valorize the accumulation of intangible assets with the introduction of technological innovations that are locally compatible and complementary with existing irreversible production factors, both tangible and intangible (Chandler, 1990, 1992; Lazonick, 1990, 1991).

In this context, considerable progress can be made by strengthening the microeconomic foundations of the relationship between growth of output and rates of generation of new technologies with notions of irreversibility, localized knowledge, localized technological change, and endogenous complementarities (Atkinson and Stiglitz, 1969; David, 1975; Stiglitz, 1987; Antonelli, 1995, 1999*a*).

All changes in expected output and factor prices induce firms to reconsider their production mix and their size. When irreversibility limits adaptive adjustment, firms start a local search in a technical space defined by the endowment of superfixed production factors. Local search leads to localized technological change in terms of increased efficiency and to the discovery of technological complementarities. Complementarities are the source of economies of scope. The advantages, both in terms of costs and revenues, of product and processes complementarities, can be reaped only when entry into adjacent markets is implemented. In so doing, firms try and extend their competitive advantage from one product to another by means of bundling strategies. Here the dynamics of localized technological change, consisting in both an induced search for higher levels of efficiency and in the induced discovery and implementation of endogenous complementarities, plays an important role and the ultimate result is the growth of the firm, both in its own product market and by means of diversification and integration, respectively, in lateral and vertical markets.

In this context, the capability to internalize the technological opportunities of the regional and technological environment—respectively, the technological districts and the technological clusters—appears to be a key factor in explaining the growth of the corporation.

2. EMPIRICAL ANALYSIS: GROWTH AND TECHNOLOGICAL CHANGE IN THE CASE OF FIAT

The automobile industry has already provided important opportunities for understanding the dynamics of technological change, the determinants of the rate and direction of technological innovations, and their economic effect (see Abernathy, 1978; Utterback, 1994). The case of Fiat, a leading car company, provides one more important opportunity.

Fiat was founded in 1899 in Turin to produce automobiles, as the name, an acronym for Fabbrica Italiana Automobili Torino, indicates. It was the fifth company to enter the young car market in Italy. The Italian automobile industry was born in 1895 when the first company was created. Another 224 firms eventually entered the industry, 224 also left it, through bankruptcy, closure, or merger and acquisition. The average life of the 224 firms is 4.5 years. Analysis of entry and exit dynamics is very interesting. Entry clusters in three periods: in the years around 1910 when 23 firms were born, in 1925 when 12 new firms were founded, and in the post-Second World War period when 15 firms started operation. Exits cluster in the same years. They peak in 1910 when 14 firms left the industry, and in 1925 and 1926 when a total of 22 firms left the market. In 1948, 6 other firms are forced to exit. Fiat contributed to the exit process through a number of relevant acquisitions: SPA in 1926, Autobianchi in 1965, Lancia in 1966, Alfa Romeo and Ferrari in 1988.[1]

---

[1] The rich data set of the Museo dell'Automobile of Turin provided a unique set of information about industrial demography in the Italian car industry.

Table 8.1. *The industrial demography of the Italian car industry, 1896–1970*

Year	Entry	Exit	No. of firms	Car production in Italy (000s)
1896	1	0	1	0.00
1897	2	0	3	0.00
1898	10	2	11	0.00
1899	10	3	18	0.00
1900	7	5	20	0.00
1901	4	6	18	0.30
1902	5	4	19	0.35
1903	5	6	18	1.31
1904	7	5	20	3.08
1905	22	4	38	8.87
1906	23	10	51	8.00
1907	7	9	49	7.00
1908	3	14	38	6.00
1909	0	11	27	6.00
1910	3	5	25	5.00
1911	4	2	27	5.28
1912	3	4	26	6.67
1913	2	5	23	6.76
1914	4	3	24	9.21
1915	0	5	19	15.42
1916	0	3	16	17.37
1917	0	3	13	25.28
1918	2	1	14	22.23
1919	5	0	19	17.90
1920	7	1	25	21.08
1921	5	3	27	15.23
1922	10	1	36	16.39
1923	5	9	32	22.82
1924	11	11	32	37.45
1925	6	11	27	49.40
1926	1	5	23	63.80
1927	2	5	20	54.30
1928	1	5	16	57.60
1929	0	1	15	55.10
1930	0	1	14	46.40
1931	0	0	14	28.40
1932	1	2	13	29.60

Table 8.1. *Continued*

Year	Entry	Exit	No. of firms	Car production in Italy (000s)
1933	0	2	11	41.70
1934	1	2	10	45.40
1935	4	4	10	50.49
1936	1	2	9	53.14
1937	1	1	9	77.71
1938	0	0	9	70.78
1939	1	0	10	68.91
1940	0	1	9	48.67
1941	1	1	9	38.80
1942	1	1	9	30.51
1943	0	0	9	21.13
1944	0	0	9	13.78
1945	0	1	8	10.29
1946	5	1	12	28.98
1947	9	4	17	43.74
1948	2	6	13	59.95
1949	3	3	13	86.05
1950	0	1	12	127.85
1951	0	0	12	145.55
1952	0	0	12	138.45
1953	0	1	11	174.31
1954	1	1	11	216.71
1955	1	3	9	268.77
1956	0	1	8	315.80
1957	0	0	8	351.82
1958	0	0	8	403.56
1959	1	1	8	500.78
1960	0	0	8	644.63
1961	0	0	8	759.14
1962	1	0	9	946.79
1963	0	0	9	1180.54
1964	1	0	10	1090.08
1965	0	0	10	1175.55
1966	0	1	9	1365.90
1967	0	1	8	1542.67
1968	0	1	7	1663.65
1969	0	3	4	1595.95
1970	0	1	3	1854.25

Analysis of the regional distribution of natality in the Italian automobile industry provides the first major element of empirical confirmation of our hypothesis. The growth of Fiat is also the result of the extraordinary technological opportunities offered by the technological district of Turin.

Seventy-three companies, over a third of the total figure of automobile companies ever started in the Italian industry were established in the Turin district.[2] This figure is larger than for the whole of Lombardy, a much larger region which ranks second with 72 firms. After these two regions the rest of the country exhibits low levels of natality: 12 companies were established in Emilia, 8 in both Tuscany and Lazio, 6 in Liguria and Veneto. The concentration of start-ups in the automobile industry in Turin is absolutely astonishing and can be interpreted only by taking into account the strong industrial and technological traditions of the town (see Table 8.2).

The birth of Fiat takes place in a proper regional context, conducive to technological entrepreneurship, with abundant capital available for risky undertakings, where abundant competencies in terms of skilled manpower were available; where an advanced school of engineering was well established, a variety of firms were testing different technological solutions in this field, and specialized suppliers were localized.

At Fiat, at the end of the first year after its foundation, 153 employees were recorded. In 1970, total employment reached 184,000 units. Production of cars and trucks increased from 300 units in 1901 to 1,854,252 in 1970.[3] Fiat quickly became competitive also in international markets, with a strong record of exports from the first years. As early as 1911, 1,314 cars were sold on international markets out of a total production of 2,474. In 1921, exports totalled 5,625 units out of 8,988 manufactured; 24,188 in 1925

---

[2] In the long term Piedmont accounts for less than 6% of the Italian population and the Turin district for less than half of this figure. The specialization index can easily be calculated at around 6.5.

[3] See the detailed empirical analyses of the growth of Fiat by Castronovo (1971), Bairati (1983), Volpato (1996), and Fauri (1996). Excellent data, in real terms, for the main economic variables such as fixed capital stock, sales, employment, wages, and value added, cars and trucks production figures, have been made available by the Fiat Archives for the years 1900–70 and they are based upon accurate in-house documents ranging from annual reports to plants' internal worksheets.

The Case of Fiat

out of 37,054. From the late 1930s and until the late 1960s Fiat sold over 30 per cent of its total production of cars and trucks on international markets. The story of Fiat is a case of rapid growth with unique characteristics in the Italian industrial context (see Table 8.3).

Irreversibility plays a major role in this case. In its first seventy years of activity, Fiat never changed its location from Turin. It moved its plants only twice. From the original plant in Corso Dante, it moved first to Lingotto and then in the early 1930s to Mirafiori. The intricate web of subcontractors, all strongly embedded in the metropolitan area of Turin and their small size was a major barrier to re-localization. The labour market, especially, acted as a major factor of irreversibility: the local supply of qualified workers was rapidly exhausted and the demand for new labour could be accommodated only via swift regional mobility of peasants with low levels of qualification and skills. Substantial limitations to extensive growth, that is, growth of output fuelled by a proportionate growth of inputs, emerged quickly. In this context, the introduction of innovations helped the company to face the continuous growth in demand for automobiles and motor vehicles driven by high levels of revenue elasticity, the diffusion of motorization, and substantial government support both in the form of protection from international competition and public demand for motor vehicles and eventually weapons.

The longitudinal analysis of the characteristics of the accumulation of technological knowledge and the technological innovations generated by the company can provide important insights into understanding the growth of Fiat.

2.1. *The accumulation of localized knowledge*

The historic analysis of the Fiat tradition in the accumulation of technological knowledge and introduction of technological innovations provides interesting insights into the evolution of dedicated organizational capabilities in the generation of technological knowledge and in the induced introduction of technological innovations. Analysis of Fiat's evolution in this field reveals the importance of three factors: (1) the close integration of the different sources of technological change; (2) the systematic attention to the active

Table 8.2. *The regional distribution of entry into the Italian car industry, 1895–1962*

Year	Lombardy	Veneto	Liguria	Emilia	Lazio	Friuli	Tuscany	Campania	Piedmont	Sicily	Abruzzi
1895	0	0	0	0	0	0	0	0	0	0	0
1896	0	1	0	0	0	0	0	0	0	0	0
1897	1	1	0	0	0	0	0	0	0	0	0
1898	4	0	0	3	0	0	0	0	2	0	0
1899	4	1	0	0	2	0	0	0	2	0	0
1900	3	0	0	0	0	0	0	0	4	0	0
1901	1	0	0	1	0	0	2	0	1	0	0
1902	1	0	0	0	0	0	0	0	2	0	0
1903	2	0	0	0	1	0	1	0	1	0	0
1904	2	0	0	0	1	0	1	0	4	1	0
1905	6	1	2	0	2	0	0	1	10	1	0
1906	7	0	1	0	1	1	1	2	7	1	0
1907	2	0	1	1	0	0	0	0	3	0	0
1908	0	0	0	0	0	0	0	0	2	0	0
1909	0	0	0	0	0	0	0	0	0	0	0
1910	2	0	0	0	0	0	0	0	0	0	0
1911	0	0	0	0	0	0	0	0	4	0	0
1912	0	0	0	0	0	0	0	0	3	0	0
1913	0	0	0	1	0	0	0	0	1	0	0
1914	0	0	0	1	0	0	0	0	3	0	0
1915	0	0	0	0	0	0	0	0	1	0	0
1916	0	0	0	0	0	0	0	0	0	0	0
1917	0	0	0	0	0	0	0	0	0	0	0
1918	2	0	0	0	0	0	0	0	0	0	0
1919	1	0	0	0	0	0	0	0	3	0	0
1920	2	0	0	0	0	0	0	0	5	0	0
1921	4	0	0	0	0	0	0	0	2	0	0
1922	6	0	0	0	0	0	1	1	1	0	0
1923	3	1	0	0	0	0	0	0	1	0	0
1924	4	0	1	0	0	1	0	0	4	0	0
1925	3	0	0	0	0	0	0	0	1	0	0
1926	0	0	0	0	0	0	0	0	1	0	0
1927	1	0	0	0	0	0	0	0	1	0	0
1928	0	0	0	0	0	0	0	0	0	0	0

Table 8.2. *Continued*

Year	Lombardy	Veneto	Liguria	Emilia	Lazio	Friuli	Tuscany	Campania	Piedmont	Sicily	Abruzzi
1929	1	0	0	0	0	0	0	0	0	0	0
1930	0	0	0	0	0	0	0	0	0	0	0
1931	0	0	0	0	0	0	0	0	0	0	0
1932	0	0	0	0	0	0	0	0	0	0	0
1933	0	0	0	0	0	0	0	0	0	0	0
1934	0	0	0	0	0	0	0	0	0	0	0
1935	2	0	1	0	0	0	0	0	0	0	0
1936	0	0	0	0	0	0	1	0	1	0	0
1937	1	0	0	0	0	0	0	0	0	0	0
1938	0	0	0	0	0	0	0	0	0	0	0
1939	0	0	0	1	0	0	0	0	0	0	0
1940	1	0	0	0	0	0	0	0	0	0	0
1941	0	0	0	0	0	0	0	0	1	0	0
1942	0	0	0	0	0	0	0	0	0	0	0
1943	0	0	0	0	0	0	0	0	0	0	0
1944	0	0	0	0	0	0	0	0	0	0	0
1945	0	0	0	0	0	0	0	0	0	0	0
1946	1	0	0	2	0	0	0	0	2	0	0
1947	3	0	0	1	0	0	0	0	1	0	0
1948	1	0	0	0	0	0	0	0	0	1	1
1949	1	0	0	0	0	0	0	0	1	0	0
1950	0	0	0	0	0	0	0	0	0	0	0
1951	0	0	0	0	0	0	0	0	0	0	0
1952	0	0	0	0	0	0	0	0	0	0	0
1953	0	0	0	0	0	0	0	0	0	0	0
1954	1	0	0	0	0	0	0	0	0	0	0
1955	0	0	0	0	0	0	0	0	0	0	0
1956	0	0	0	0	0	0	0	0	0	0	0
1957	0	0	0	0	0	0	0	0	0	0	0
1958	0	0	0	0	0	0	0	0	0	0	0
1959	0	0	0	0	0	0	0	0	1	0	0
1960	0	0	0	0	0	0	0	0	0	0	0
1961	0	0	0	0	0	0	0	0	0	0	0
1962	0	0	0	1	0	0	1	0	0	0	0

Table 8.3. *Fiat: basic statistics*

Year	Employees (000s)	Value added (millions)	Sales (millions)	Capital (millions)	Patents
1900	0.15	0.000	0.000	2.130	0
1901	0.16	1.388	1.187	2.102	1
1902	0.18	1.635	2.466	2.174	4
1903	0.24	2.207	4.464	2.273	0
1904	0.50	3.175	7.977	3.006	4
1905	1.13	8.254	16.700	48.000	4
1906	2.50	17.018	36.031	50.094	10
1907	2.74	41.041	69.468	50.406	1
1908	2.42	27.212	71.599	54.806	1
1909	2.53	29.581	80.918	53.130	3
1910	2.75	34.797	98.523	54.163	7
1911	3.00	44.676	109.342	50.566	10
1912	3.23	52.949	151.802	54.570	8
1913	3.48	57.708	171.999	40.956	6
1914	3.89	63.222	192.889	40.101	3
1915	5.10	78.819	234.903	32.905	1
1916	10.41	107.654	277.793	21.234	4
1917	16.54	144.475	522.168	33.179	15
1918	17.30	145.672	527.405	15.571	17
1919	13.91	143.612	467.154	10.942	20
1920	15.02	110.435	357.120	11.679	0
1921	10.05	94.092	244.904	11.502	32
1922	8.01	98.571	285.178	13.217	15
1923	17.71	78.367	256.650	14.936	25
1924	25.50	134.168	321.536	17.256	25
1925	30.65	202.114	458.749	23.133	47
1926	30.08	230.909	661.109	24.308	16
1927	26.09	219.499	693.954	28.422	13
1928	31.51	238.733	683.336	30.920	21
1929	28.33	278.692	694.847	38.609	20
1930	23.73	264.727	796.068	39.818	53
1931	22.85	245.896	729.964	42.352	30
1932	23.89	215.275	1,019.203	37.959	38
1933	26.18	231.075	1,273.124	33.466	39
1934	27.47	268.665	1,497.511	26.135	39
1935	38.49	307.323	1,735.425	16.505	33
1936	44.15	368.733	1,763.268	8.259	6

Table 8.3. *Continued*

Year	Employees (000s)	Value added (millions)	Sales (millions)	Capital (millions)	Patents
1937	50.15	395.723	1,657.052	12.000	29
1938	51.86	417.393	1,458.984	15.000	46
1939	55.67	440.189	1,523.996	40.000	58
1940	61.13	684.369	2,172.676	60.000	34
1941	64.41	929.127	2,819.994	120.000	29
1942	64.22	1,257.072	3,489.216	180.000	34
1943	63.87	1,366.230	3,737.531	210.000	20
1944	62.91	1,650.307	3,310.645	243.651	19
1945	64.38	2,437.875	3,399.682	86.000	14
1946	64.52	260.461	235.610	45.420	27
1947	68.12	577.645	519.861	51.414	24
1948	66.36	305.334	684.594	333.296	7
1949	71.21	487.048	1,076.101	343.559	28
1950	72.67	754.447	1,407.354	480.401	27
1951	72.04	942.433	2,132.934	660.663	31
1952	70.00	1,109.169	2,039.361	771.245	50
1953	71.11	1,093.484	2,540.362	1,303.894	41
1954	71.30	1,293.492	3,086.625	1,488.572	31
1955	74.89	1,469.713	3,526.685	1,765.060	33
1956	77.32	1,684.005	3,920.903	2,077.003	34
1957	80.42	1,838.784	4,234.367	2,385.497	30
1958	79.93	1,914.231	4,332.295	2,855.059	23
1959	85.12	2,300.000	4,674.154	3,191.644	28
1960	92.89	2,707.880	5,655.110	3,469.081	47
1961	107.67	1,871.410	5,849.600	3,792.102	34
1962	119.84	2,686.938	6,883.030	4,483.797	58
1963	126.32	3,076.370	8,320.400	5,196.856	41
1964	124.34	3,649.954	9,353.320	5,855.527	44
1965	123.11	3,724.415	8,436.000	6,324.862	38
1966	134.59	3,965.859	8,967.000	6,686.989	43
1967	144.50	5,818.097	11,591.450	7,259.421	51
1968	158.45	6,788.849	13,217.580	8,359.992	54
1969	170.88	6,715.323	14,725.050	9,756.061	52
1970	184.81	7,425.323	15,133.500	10,557.257	67

Note: All figures are in millions of real liras. Employment is in thousands.

acquisition and implementation of a wide spectrum of different and yet complementary sources of knowledge, and (3) the interaction between business and technological strategies. Let us consider these in turn.[4]

At the firm level, technological change is the result of three distinctive processes: (1) the introduction of endogenous technological innovations; (2) the imitation of technological innovations introduced by competitors; (3) the purchase of capital and intermediary goods embodying technological changes introduced in upstream industries. These three sources have been very important in the Fiat case. Their sequential integration has been the result of an active strategy. Especially in the first years of activity, imitation and the purchase of technological innovations embodied in capital and intermediary goods played a major role. For at least twenty years, introduction of basic innovations such as welding, sheet steel rolling, and die casting was mainly the result of imitating US manufacturers.[5] Endogenous innovation has been gradually introduced as the result of a long-term strategy of building internal competencies and implementing actual innovation routines.

The charismatic role of Dante Giacosa, the Chief Engineer of the Company for over forty years and eventually an influential member of the Board, was fundamental in creating an internal innovation routine and a basic competence in accumulating technological knowledge and eventually generating technological innovations at Fiat. The distinctive character of his approach to technology management consisted in the proactive integration of the interfaces

---

[4] Much evidence presented in this section is based on interviews with former managers and workers who were either retired or nearing retirement. Important oral evidence has been provided by many interviews with Dr G. De Giorgis, for many years the head of the Ufficio Brevetti at Fiat, Dr M. Actis, Dr E. Cordiano, Prof. G. Berta, at the Archivio Storico of Fiat, Prof. U. L. Businaro, for many years Chief Scientist of Fiat, and Prof. Cesare Annibaldi. Important hints are also found in the minutes of the meetings of the Board of Directors, available since foundation and until the 1930s (see the Fiat Archives).

[5] Since foundation and during the 1920s, technological and organizational changes introduced by Ford were kept under close scrutiny by Fiat and quickly imitated. Fiat also relied heavily on the suppliers of Ford with an interesting *triangular* process of technological imitation, whereby innovations introduced by Ford but manufactured by its suppliers were eventually imitated by Fiat via the purchase of the pertinent capital goods.

The Case of Fiat

between four different forms of knowledge: internal and external, tacit and codified knowledge (Bassignana, 2000).

Internal tacit knowledge was valorized through the direct involvement of shop-floor workers who were invited to suggest new organizational devices and technological suggestions for which special incentives were provided. Vocational training was provided internally by the famous 'Scuola Allievi Fiat' where trainers were often former blue-collar workers so as to create an internal direct flow of tacit competence from old expert workers to new young ones.

External tacit knowledge was actively searched for and assimilated by means of the recruitment of skilled manpower from competitors and specialized suppliers. External tacit knowledge has been much acquired as a result of the aggressive strategies of external growth, both horizontal and vertical. In this context, a special role has been played by the systematic strategy of vertical growth along the upstream *filière* of producers of intermediary inputs, often realized by means of merger and acquisition of smaller firms which brought under the control of Fiat a large portion of the Piedmontese mechanical engineering industry. The acquisition of new firms in fact enlarged the scope for internal mobility of skilled manpower and favoured the circulation of technological information among engineers and technical staff which gradually extended their expertise and technical competence to the full vertically integrated production process, from metal products to transportation equipment.

Access to external codified knowledge and eventual recombination was actively practised with intense relationships with local universities and especially the Politecnico of Turin with which a long-lasting cooperation has been taking place with the direct participation of many academics, often acting as consultants and part-time employees, in the research projects conducted by the firm, the selective hiring of the best Italian graduates from engineering school, and the systematic technological borrowing from foreign-specialized suppliers and competitors. Close relations with US machine tools companies in the first decades of the twentieth century have been documented by Volpato (1996) who provides evidence about the important role of foreign suppliers of embodied technological innovations, especially in upstream activities. The purchase of US patents and licences was also practised and systematic reverse engineering took place. Recombination of foreign technology with domestic traditions emerged as a distinctive source of technological knowledge.

Finally internal codified knowledge was elaborated and implemented with a single centralized R & D unit, responsible for the management of the interfaces between the different forms of knowledge so far acquired. Fiat was able to establish a reputation as a strong technologically minded company with a tradition of technological excellence and a strong technological base which attracted the best engineers in the country for many decades. The traditional career path, in fact, included the direct involvement of graduate engineers at the shop floor for many years after recruitment and a dual wage ladder, such that seniority on the shop floor was often highly rewarded. The mingling of qualified engineers and blue-collar workers on the shop floor was considered a direct source of valuable technological competence and suggestions.

It seems important to stress here that this technology management tradition is to a large extent the final result of a historic process shaped by the 'accidental' leadership of the charismatic Dr Dante Giacosa. Dante Giacosa played a key role in the gradual institutionalization of innovative routines within the company and mastered the transition from a situation where external knowledge was the main source of technological innovations to a deliberate technological strategy, based on internal resources. On the other hand, Dante Giacosa resisted the trend towards the identification of a formal R & D unit as the single centre of accumulation of technological innovation and kept open a variety of ways to accumulate and implement the technological competence of the company. Formal research and development activities remained at very low levels.[6] The active recombination of external codified knowledge, mainly absorbed by means of the use of academic personnel on a consultancy basis, the access to external tacit knowledge via mobility on external labour markets and the acquisition of firms in upstream activities, the active involvement of internal skilled manpower, were the main sources of technological knowledge.

---

[6] The assessment of proper R & D expenditures is made difficult by the lack of data. The notion itself of research and development activities dates from the early 1960s and has been used by Italian companies only since the early 1970s.

The Case of Fiat

In this context, the 'wage and industrial relations strategy' of Fiat with respect to qualified blue-collar workers, often assigned to maintenance duties, and foremen seems relevant. A large body of empirical evidence confirms that wages for such personnel have traditionally been high in Fiat for almost all the period considered, with a premium above average wages in the car industry in Turin which has never been lower than 45 to 50 per cent.[7] Secondly and most important, long-term employment security has been used by Fiat as a tool to secure the dedicated commitment of a selected number of blue-collar workers. Fiat was able to keep its promise of employment security for qualified blue-collar workers even in the depressed conditions of the 1930s. Unionization paralleled this process with high levels of participation by skilled workers and much lower levels for other categories. For a large part of the period considered, unionization was also a factor that integrated workers in the organization of the company. Finally, upward intergenerational mobility in internal markets has been the third factor securing the loyalty and commitment of creative and selected workers. Not only could qualified workers progress in terms of wages and hierarchy, with substantial opportunities for job enrichment and job enlargement, but employment in clerical positions was sought and actually achieved for their children.

Both the levels and the typology of tasks and job assignments selected suggest that Fiat has been practising efficiency wage when and where the accumulation of tacit knowledge and competence was considered strategic. Dynamic efficiency wages in Fiat enhanced loyalty and commitment and stimulated practitioners, mainly foremen, to develop informal relations and better collective work, sharing information and accelerating the emergence of tacit knowledge.

The active participation of a qualified workforce in implementing learning processes made it possible to accumulate and better valorize tacit knowledge and experience, enabling the proper evaluation of the specific context of action, and enhancing the matching

---

[7] Musso (1980: 181) documents that wages in Fiat were far higher than industrial and regional averages: respectively 178% of the average in 1914, 191% in 1918, 173% in 1919, 176% in 1920, 145% in 1921 in the metropolitan area of Turin, itself higher than the average in Piedmont, in turn the highest in Italy.

between the availability of new codified knowledge and experience within the firm. The rates of implementation of know-how, know-where, and know-when, in the Fiat experience, seemed to rely systematically on levels of participation of the skilled workforce in both production and decision-making. Especially in the first twenty years of the twentieth century, workers were actively solicited to contribute to the implementation of the production process and formal rewards were granted to new suggestions. A formal procedure had been established for the scrutiny of new bottom–up ideas which were first considered by foremen and eventually transmitted to a group of experts and engineers for final evaluation.

Effective internal labour markets which favoured the upgrading of competent employees within the firm have also been an important complementary tool to accelerate rates of accumulation of experience and tacit knowledge in Fiat. In fact, strong vertical mobility for blue-collar workers kept competent labour within the firm and acted as a powerful incentive to stimulate the participation of the workforce in learning processes.

In sum, dynamic efficiency wages in Fiat seem to have stimulated inductive processes of learning by doing and learning by using among the workforce, feeding the accumulation of localized knowledge, through encouraging both the bottom–up process of accumulation of competence and innovative capability and the top–down process of adaptation of new codified knowledge to the idiosyncratic context of the firm. Efficiency wages facilitated the processes of 'translation' of tacit knowledge, acquired by means of learning processes within the workforce, into codified knowledge, and vice versa. The Fiat evidence confirms that efficiency wages and internal labour markets had a powerful effect, accelerating the blending of internal and external knowledge and its integration with the organizational knowledge on which the introduction of localized innovation rested.

Technological communication, in managing the interfaces between diverse and yet complementary sources of knowledge as well as diverse and complementary types of knowledge, has played a key role as the integrating device. Analysis of the evolution of the managerial hierarchy reveals constant attention to task units created to implement and keep open the interactive flow of information both vertically within the company from the shop floor and through the

The Case of Fiat

different layers of the managerial structure and horizontally with respect to the increasing range of products and related production activities.

The interviews confirm the strong awareness of the Fiat management that the production of car and other transportation equipment implied not only the active coordination of different forms of knowledge—internal/external/codified/tacit—but also of different types of knowledge. The production of new knowledge in transportation vehicles impinged upon the creative command of a wide array of different specific technologies, each with their own research tradition, disciplinary characteristics, and general scientific principles. Such a spectrum ranges from mechanical engineering, to electrical engineering, chemistry, physics, oleodynamics, metallurgy. Resistance to the introduction of the multi-divisional form was based upon the strong vision of the need to preserve the integrating role of a common technological competence which was nevertheless fed by a variety of applications.

Technological knowledge in transportation equipment was, in other words, viewed as the result of a multi-technological undertaking where induced technological convergence played a central role. The generation of each piece of new technological knowledge in this activity was heavily influenced by the outcome of a variety of scientific and technological advances made in a wide range of distinct scientific and technological fields. Their knowledge base, however, was distinct and drew upon research traditions which were highly specialized and separate. In this context, the conditions of technological communication became key factors in the generation of new knowledge. Technological communication, among a variety and complexity of details and applications in which new knowledge was embedded, was viewed as the basic requirement to generate new technological knowledge. Hence the need for dedicated managerial competence in implementing all possible interfaces, ensuring, on the one hand, the autonomous implementation of each and yet, on the other, their coherent and complementary development.

The systemic nature of such knowledge, on the one hand, made technological communication central and, on the other, increasingly accounted for the emerging cumulability, compatibility, complementarity, and interdependence of these diverse bits of technological knowledge embedded in the specific and idiosyncratic features of each learning unit within the firm. The recombination and generalization of such partial and incomplete forms of applied knowledge provided new ways and resources to generate additional generic knowledge, germane to subsets of more specific and productive applications. Hence, because of cumulability, compatibility, and complementarity, technological communication became more and more important for further advances to take place.

In sum, attention to the mechanisms of technological communication among diverse forms and diverse types of technological knowledge has gradually emerged as the distinctive feature of technology management in Fiat, while demand pulled the company to cope with increasing levels and an increasing variety of output. The Fiat case can be considered from this viewpoint as an early example of a company based upon internal communication mechanisms.

Finally, the evidence suggests that technological strategies and business strategies at large were kept in close interaction and had a strong reciprocal influence. The basic intuition of Dante Giacosa, which became the Fiat approach, was the understanding of the key role of technological change as a tool to support business strategies. This strategy has been implemented in three ways: (1) through the search and eventual introduction of process innovations in the automotive sector. The fast rates of increase of demand for automobiles and the constraints to extensive growth pushed the technological strategies of the company towards the introduction of productivity-enhancing innovations which could sustain the growth of the company. Technological strategy here was consistent with the search for reductions in costs and increased flows of outputs. (2) The direction of technological change was deliberately characterized by the search for *sequential cumulability*[8] with the existing endowment of the company in terms of fixed capital and product characteristics. Research directions and technological opportunities were selected according to the complementarity and interoperability of eventual innovations with existing production lines and other inputs, in order to reduce adjustment costs and valorize the scope for further use of sunk factors. (3) Finally the strategy involved the introduction of product innovations coherent with a strategy of

---

[8] It may be interesting to know that Dante Giacosa was proud to have been able to implement and diversify the applications of a single basic engine introduced in the early 1920s to a variety of automobiles and other machinery. The same engine was actually dismissed only in the late 1960s.

The Case of Fiat

related diversification towards new segments of the transportation industry as a whole, including trucks, agricultural machinery, and eventually aeroplanes, marine engines, and weapons. Technological strategy here was clearly aimed at coping with the growth of the demand for new products, often driven by relevant public procurement. New products, however, were introduced under the constraint of the systematic extension of the scope of application of components and related production processes already in place for automotive activities.

This is especially evident when analysing the technical continuity and contiguity of an array of components used in the different production processes from the brakes, to gears, from the electrical and control systems to the engines themselves. Economies of scope here are the endogenous result of an intentional strategy of extension and valorization of the opportunities for applications of a single, and yet evolving, endowment of fixed capital and competence. The evidence suggests that, in this case, latent technological complementarities have first been successfully made explicit and then internalized, so as to lead to major economies of scope and eventual growth by diversification with entry in a number of adjacent product markets.

2.2. *Patent data and the rate and direction of endogenous technological change*

The actual rate and direction of technological innovations introduced by Fiat in the period 1900 to 1970 is difficult to observe and measure. A proxy can be provided by patents delivered to Fiat by the Italian Patent Office. Patent data for Fiat have been collected since its foundation in 1900 up until 1970, with the help of the office of the chief scientist of the company.[9] Table 8.3 shows the number of patents delivered to Fiat. Each patent has been assigned to the date of first submission to the Italian Patent Office; hence actual

---

[9] All the patents for the period considered were still kept by the Ufficio Brevetti in different files. They have been collected on purpose for this study, reorganized, and eventually transferred to the Archivio Storico. All patents are recorded and are now available at the Fiat Archives.

delivery took place with varying time delays. The submission date coincides with the decision of the chief scientist of the company to pursue the patenting of that innovation. Innovations for which patents were not assigned are not counted, nor are innovations for which patents were not requested.

Patent counting has often been used in the applied economics of innovation and new technology as a reliable empirical indicator for the rate and direction of technological innovation. Our context of analysis seems even more reliable than usual. Important biases, such as the high variance in the actual technological and hence economic value of patents and the diversity between the identity of the patent holder and the specific company or laboratory where the innovation originated are far less relevant. Patent counting at the firm level, however, suffers from one important limitation: it can provide information only about endogenous innovations. This is not the single type of technological change introduced at the company level. Sheer imitation of technological innovations introduced by other firms and embodied technological change purchased with new capital goods cannot be observed by patent data and yet play an important role in assessing the overall rate of technological change.

Patent data, in this case, have originated from one company with high levels of time consistency with respect to their content: the same persons have been in charge of the decision-making about patents for over thirty years. The screening of the different technological innovations elaborated by the company and its evaluation has been directly assessed by a small internal team[10] which remained in charge for a long time period with just a few changes.

The distribution of patents in the period considered reveals a significant growth from 1 patent in 1900 to 67 in 1970 with important fluctuations. A first local peak is reached in the years around 1910 with a flow of 10 patents per year. In the 1930s, the flow of patents reaches a new local maximum of over 50 patents per year which is maintained until the Second World War. In the post-Second World War period, a new stage of growth is implemented and patents per year increase from an average level of 20 patents in the late 1940s to the maximum of 1970. Innovative effort, as measured by the ratio of patents to employees, shows a substantial decline. This

---

[10] The so-called Ufficio Brevetti.

The Case of Fiat

pattern lasts for the whole period considered, but for three important upward fluctuations around 1910, in the early 1920s, and in the second part of the 1930s.

Inspection of the files containing the technical description of each innovation, eventually patented, reveals a technological content which can be characterized as typical product/process innovations. The early patents reflect the key role of technological breakthroughs such as developments in welding, sheet steel rolling, and die casting. Since the 1920s, the patents consist mainly of technological innovations in specific components of automobiles and eventually other transportation equipment such as trucks, agricultural machinery, aeroplanes, weapons. In the later period, the technological innovations considered mainly concern new engines, brakes, transmission systems, electrical systems for cars and other transportation equipment.

New models are not covered by the patents considered. A large number of new production processes are also considered and they are all directly related to the innovations in components already considered. Patent counting in our case seems to identify an important, albeit specific, area of technological innovation such as the introduction of new technologies which affect both the overall performance of the manufacture of transportation equipment considered, whether cars, trucks, or agricultural machinery, and the hedonic price of the final products.

Although the qualitative assessment of patents is a complex task, the general evidence emerging from an analysis of the technical contents suggests that rarely do patents delivered to Fiat cover radical technological innovations with revolutionary effects. Rather, the technological novelty of patents would appear to match the notion of incremental innovations with a strong cumulative character. From this viewpoint, the record of patents tells a story of incremental technological innovations that have been introduced systematically and sequentially around well-defined technological paths consisting of (1) the identification of specific functions of automobiles and their technological upgrading; (2) the specification of an increasing array of automobile components and their production process; (3) the search for new materials to be used in the production of transportation equipment; (4) new processes to manufacture them; and (5) the implementation of latent technological economies of scope, that is, the application to other products, ranging from trucks to weapons (mainly armoured vehicles and aeroplanes), of a number of technological novelties. Frequently, new patents make explicit reference to previous ones and new technologies are presented as developments and specifications of technological innovations already patented. Recombination and learning seem to be the main sources of this technological knowledge and confirm the important role of the accumulation of competence and skills as the driving force behind the path to technological advance followed by Fiat. This confirms the strong cumulative nature of technological innovation in Fiat, with effects which could easily be interpreted as the outcome of the dynamics of learning to learn in a well-defined and localized technical space.

Analysis of the flow of patents across the seventy years considered makes it possible to identify three relevant dynamic processes at play: the role of systemic unbalances; the recursive feedbacks between core and diversified technological activities; and the interaction between product and process innovations leading to the specification of an increasing array of self-contained components. Although these three processes have been in place for the whole period considered, each of them was especially relevant in one of three distinct sub-periods: the first in the years from 1900 up until the late 1920s; the second in the 1930s; and the third in the post-Second World War period.

The Fiat case in the first three decades provides evidence about the important role of the systemic character of technological change. Automobiles were (and still are) complex systems where changes in each component and functional aspect have direct consequences on the others. Many technological innovations introduced by Fiat in these years seemed to respond to this internal systemic mechanism, driven by the rapid growth of demand and the related search for efficiency. The flow of patented innovations documents the pervasive role of this internal process: the introduction of changes in engines exerted a clear effect both in terms of inducement and capability on the introduction of innovations in brakes systems. The introduction of innovations in electrical apparatus had direct effects on transmission equipment. The search for new metallurgical technologies led to reductions in weight which drove technological changes in the control of the speed and the consumption of gasoline and so

The Case of Fiat

on. At this time, external knowledge, both embodied in advanced machine tools, often imported from the USA, and disembodied in the form of licences—again from US firms, played a major role, together with the craft expertise of highly qualified senior foremen and repairmen. The 'innovative locus' within the company was clearly situated behind the lines of the main production process which is still characterized by batch production.

Analysis of the longitudinal records of technological innovations introduced in Fiat during these years suggests that a strong cumulative inducement mechanism has been at work. More specifically, a distinction can be introduced between vertical and horizontal cumulativeness. In the former case, many new technologies were clearly standing on the shoulder of previous innovations in terms of direct accumulation of technological knowledge. In the latter case, attention is focused on the mechanism by means of which the introduction of innovations in each specific function of the complex technological system of an automobile appeared to be the result of the inducement and focusing mechanisms activated by the introduction of innovations in complementary and interrelated parts of the same system. From this viewpoint, analysis of the Fiat case confirms the relevance of systemic approaches, advocated by Paul David (1987 and 1992), in understanding the forces at play and assessing the direction and rate of technological change.

Analysis of technological changes at Fiat during the 1930s confirms the insights of Chandler (1990, 1992) about the key role of the interaction between economies of scale and economies of scope in explaining the growth of firms. Economies of scope, however, emerge as the intentional outcome of a research, production, and commercial strategy aimed at the search, implementation, and valorization of all latent complementarities, both in the production of goods, in their market distribution, and in the generation of new technological knowledge.

The Depression and stagnating demand for cars forced the company to try and widen the scope of application of its technological knowledge: strong emerging demand for trucks, agricultural machinery, and military equipment was again the push factor. This was reinforced, however, by important feedbacks. The interaction and sequence between centrifugal-specializing innovations[11] and centripetal-diversifying ones are clearly among the leading forces behind the flow of innovations introduced by Fiat during these years. In the 1930s, technological knowledge originally accumulated in the production of cars systematically spread to other transportation equipment with significant technological economies of scope. At this time, the role of internal knowledge residing in the engineering and design departments—now the main innovative locus—played a major role, although its character was mostly tacit; external codified knowledge was also important and it was mainly provided by academic consultants. Internal tacit knowledge was mainly active in the reverse feedbacks from diversified activities to core ones specializing in car production.

Inspection of patent files and complementary interviews with the director of the Fiat patent office suggests, in fact, that a cumulative core–periphery interaction between technological changes in the production of cars and related components and technological changes in other transportation equipment had been taking place. Such interaction was clearly two-way, in that not only were new technologies originally developed in car production and for car production actively applied to other transportation equipment, but also the reverse was true. Applications of car-dedicated technologies in the production of components for trucks or agricultural machinery led to incremental and cumulative technological changes which could eventually be applied to the production of cars and their components after minor incremental changes. This process was especially relevant in brake and transmission technology, where creative adaptations of original car-dedicated innovations to trucks led to specific innovations which were eventually applied in car production.

Finally, the sequence of patents analysed suggests that much technological innovation in the post-Second World War period (from the 1940s up until the 1960s) originated through keen attention to the interface between product innovations and new process technologies: the introduction of new families of products was often

---

[11] It can be claimed that the eventual break-up of the Fiat group, begun in the 1970s, into a holding company controlling a variety of companies specializing in diverse transportation equipments and components finds its primary source in the direction of technological changes introduced in the years analysed.

followed by an array of technological changes in the actual manufacturing process. At this time, internal codified knowledge played a major role, together with more systematic institutional relations with local universities; once more, however, internal tacit knowledge, both on the shop floor and in the engineering departments, played a major role. It is worth noting that the 'innovative locus' in this phase shifted away from the actual main plants—now fully converted to mass production—and was mainly found, upstream, in the mechanical units which produced capital goods for the assembly line or in newly acquired small companies active in the emerging plastic industry.

This evidence confirms the well known product-process innovation sequence first assessed by Jim Utterback in analysing the dynamics of technological change (Utterback, 1994). The rapid rise in demand in the first two decades of the post-war period pushed the introduction of technological innovations about specific functions of the car and other transportation equipment to feed the flow of new models - as such they were clearly product innovations. The continuous introduction of new models and related technological search paved the way to the identification of new self-contained components which eventually became products in their own right and with respect to which subsequent process innovations have been introduced. The relationship between product and process innovations is in other words quite complicated in that technological changes in car—product innovations—lead to significant changes in the production of cars, as such process innovations, which however lead to the identification of new products—the components—whose production process eventually became an area of technological change. The sequence in other words seems to be product innovations which lead to process innovations which lead to new product innovations in a spiralling self-reinforcing and cumulative process.[12]

---

[12] The actual development of the car components industries, a process which can be dated in Italy from the 1950s, as areas of identifiable Marshallian business and technological activity, distinct and separate from the automobile industry itself, can be considered from this viewpoint to be the result of this very process of sequential interaction between product and process innovations in the increasing identification of self-contained components.

2.3. Statistical analysis

The case of Fiat in the years 1900–70 seems especially pertinent to test our microeconomic model of induced and localized technological change. The evidence, in fact, confirms that in this case the basic technical, market, and technological conditions conducive to the induced generation of localized technological change, apply.

First, irreversibility is very high in a complex production process, such as a highly vertically integrated transportation equipment manufacturing company, where final output is the result of a variety of sub-routines, each of which has high levels of indivisibilities, interrelatedness, and complementarities. Changes in output by means of changes in irreversible inputs are especially expensive because of the need to maintain the necessary complementarity among all the specialized production units involved.

Secondly, the dynamics of technological innovation shows that Fiat had a major technological opportunity based upon a systematic strategy of related diversification and cumulative technological change. The company widened the scope of application of each single invention from cars to trucks, to agricultural machinery, weapons, aeroplanes, and naval engines and vice versa in a spiralling bottom–up process of technological diversification and vertical integration, enrichment and generalization of the basic technological competence.

Thirdly, the characteristics of the technological district of Turin played a major role in terms of the quality of local markets for skills, professional services, and advanced components, and most important, the strength of the academic community. The internalization of such externalities and access to the local pool of collective knowledge have been the driving forces behind the growth of the company.

Finally, the generation of technological knowledge in engineering relied heavily, as revealed by interviews with the company's chief scientist and confirmed by qualitative analysis of the technical description of patents, on learning processes strongly associated with the actual techniques in place at each point in time. In this process, the dedicated efforts of a competent labour force played a major role in expanding the width and depth of the stock of technological competence within a well-defined corridor of evolving techniques.

The combination of these specific conditions leads to a case where the incentives to cope with factor and product market dynamics by means of the introduction of technological changes, within a localized technical space, are very high, both because of high levels of irreversibility and hence switching costs and due to significant opportunities to generate technological innovations, based upon bottom–up technological opportunities.[13]

This set of conditions provides an excellent opportunity to test the model of localized technological change so far elaborated. The model of localized technological change elaborated in Chapter 2 and reconsidered specifically in Section 1 provides the basic guidelines on which to implement the basic hypotheses for the empirical analysis.

According to our hypothesis, rapid growth has been an important determinant of the flow of endogenous technological changes introduced by Fiat. The company, constrained by the substantial irreversibility of the production process, in an effort to cope with the rapid growth of the Italian market for cars and other transportation equipment and with the fast increase in wages, has been induced to generate localized technological changes in order to increase the productivity and reliability of products and its competitive advantage in the market-place, drawing on the substantial knowledge externalities available both in the technological district of Turin and in the emerging mechanical engineering technological cluster. So far, we expect to test the direct positive effect of the growth in employment on the amount of innovative activity and, in turn, of the pace of endogenous technological change on output growth.

The test of this hypothesis is organized in two nested steps. First, we test the hypothesis that innovative activity, as measured by patents, is a consequence rather than a cause of growth, under the constraint of superfixed production factors. On this basis, we can then test the hypothesis that patents, now an endogenous variable, have played a key role in explaining the growth of output of the company. This sequence of nested tests makes it possible to distinguish our endogenous model of innovation, induced by demand

---

[13] An important distinction is introduced here between 'top–down' or generic technological opportunities spilling from scientific discoveries, and 'bottom–up' technological opportunities engendered by learning processes in limited technical spaces defined in terms of factor intensities, location, skills of labour force typology of existing fixed capital and commercial practices (Antonelli, 1999a).

growth, under the constraint of superfixed production factors, from the traditional model, according to which exogenous technological change explains output growth.

Given the time series nature of the data available, a simple Granger causality test of the percentage change in employment on the percentage change in patent flows can prove interesting and convincing. The hypothesis is that employment growth Granger-causes patent growth. This hypothesis is the reverse of the standard model in which patents increase productivity, which leads to job growth.

The Granger causality test has been performed with two, three, and four-year lags. The hypothesis that causality runs between patents and growth, that is, that the percentage increase of patents does cause the percentage increase of employment (the dependent variable) can be rejected with an F-statistic of 3.61193 (and a probability of 0.03246) with a two-year time lag; an F-statistic of 3.80092 (and a probability of 0.01435) with a three-year time lag; an F-statistic of 3.03247 (and a probability of 0.02413). The hypothesis that the causality runs between growth and patents, that is, that the percentage increase of employment does cause the percentage increase of patents (the dependent variable) cannot be rejected with an F-statistic of 0.09701 (and a probability of 0.90768) with a two-year time lag; an F-statistic of 0.26764 (and a probability of 0.84848) with a three-year time lag; an F-statistic of 0.62193 (and a probability of 0.64864).[14]

Our second nested hypothesis argues that the flow of (the now) endogenous technological innovations, measured by the flow of patents, has been instrumental in explaining the evolution of output through its positive effect on total factor productivity. The basic argument here is that the endogenous generation of localized technological innovations has been used by the company to complement, that is, partially substitute for, the rate of growth of inputs necessary to cope with the fast growth of demand, under the constraint of major superfixed production factors. Hence we expect to measure a clear and strong positive effect of patents on output, within the context of a standard technological production function (Griliches, 1979, 1984).

---

[14] The comments of an anonymous referee are acknowledged.

The econometric specification is as follows:

$$\ln Y = a + b(\ln K) + c(\ln L) + d(\ln P) + f(\text{Time}) + e \quad (8.1)$$

where Y measures value added in real terms; K measures fixed capital in real terms; L measures employment in units; P measures patents in units; Time measures time flow; and e is the expected error term.[15]

The econometric estimate has, again, been conducted on the full time series, from 1900 to 1970, with the exception of the years 1916–19 and 1943–47, because of evident alterations in the time trends due to the war periods. The results of the OLS test are satisfactory in terms of the explanatory power of the regression equation:

$$\ln Y = 3.414 + 0.349 \, (\ln K) + 0.616 \, (\ln L) + 0.075 \, (\ln P)$$
$$(5.771) \qquad (6.257) \qquad (2.931)$$
$$+ 0.019 \, (\text{Time})$$
$$(4.103) \qquad\qquad\qquad\qquad (8.2)$$

$$R^2 = 0.988; F(5–57) = 1148.834; DW = 1.12.$$

(t statistics in parentheses)

These results show that the output elasticity of the patent flow reaches the heights of 0.075. It should be evident that the magnitude of this elasticity is especially relevant. The strong and positive effect of the flow of patents confirms that the accumulation of skills and competencies have played a major role in the explanation of the outstanding growth of Fiat since its foundation.

In order to handle properly the risks of simultaneity between output and endogenous innovative activity and to appreciate the time distribution of the effects, a lagged specification of P has also been introduced. The results confirm that the moving average of P with

---

[15] It is well known that ordinary least squares estimates of time series data may incur spurious relationships. This time series, however, is microeconomic and exhibits non-irrelevant discontinuities. In these conditions, the insertion of a time variable seems sufficient to cope with the trend effects (see Taylor and Dixon, 1997).

a three-year time span (P3) exerts the strongest effect:

$$\ln Y = 1.728 + 0.298 \, (\ln K) + 0.712 \, (\ln L) + 0.187 \, (\ln P3)$$
$$(6.664) \quad\quad (13.474) \quad\quad (2.272)$$
$$+ 0.019 \, (\text{Time})$$
$$(2.961) \tag{8.3}$$

$R^2 = 0.990; F(5-61) = 1380.696; DW = 1.12.$

(t statistics in parentheses)

Overall assessment of the results shows that Fiat has benefited from substantial *dynamic increasing returns* which find in the flow of technological innovation and in the related accumulation of competence a much stronger explanation than standard assumptions about technical economies of scale based on the sheer increase of size of a given technology.

The quality of these results is confirmed when the results of the estimates, for the same data set and time span, of a simple Cobb-Douglas production function are compared:

$$(\ln Y) = 2.887 + 0.333 \, (\ln K) + 0.822 \, (\ln L) + 0.019 \, (\text{Time})$$
$$(6.647) \quad\quad (16.617) \quad\quad (3.305)$$
$$\tag{8.4}$$

$R^2 = 0.986; F(5-61) = 1841.902; DW = 1.019$

(t statistics in parentheses)

The estimated parameters of the marginal productivity of capital and labour are robust and significant at more than 99 per cent probabilities. The results of equation (8.4) would suggest that significant increasing returns are at work: the sum of the estimated parameters of the marginal productivity of capital and labour is well above unity. Our model and the estimates so far gathered, however, suggest that a different story is at work. The apparent increasing returns estimated in equation (8.4) are determined by the omission of a key variable: the flow of localized technological innovations introduced by Fiat in these years which are themselves the outcome and the cause of the growth in size of the company. All this becomes clear when the residuals of the estimates of equation (8.4) are computed and regressed on the three-year moving average of the patents delivered to Fiat in the years considered. The OLS estimates now read as follow:

$$\text{RESIDUALS} = -0.234 + 0.59 \, (\ln P3) \tag{8.5}$$
$$(2.146)$$

$$R^2 = 0.126; F(1-57) = 18.527;$$

(t statistics in parentheses)

where RESIDUALS are the residuals of equation (8.4) and (P3) the moving average with a three-year span of the patents delivered to the company.

These results confirm that the variable introduced to catch the effects of technological innovations and technological knowledge adds explanatory power to the regression equation without evident effects of multicollinearity.

The results of the statistical analysis are important because they lend new support to the Penrosian and Chandlerian conjecture, albeit with some qualifications. The combined interpretation of the results of the Granger causality test which confirms that innovative efforts, as measured by patents, are the dependent variable and employment growth the independent variable, and of the results of equations (8.1–8.5) where innovative efforts are an independent variable explaining, respectively, output and hence total factor productivity (and labour productivity), under the control of capital intensity, confirms the effects of growth on productivity, via the induced introduction of innovations.

3. CONCLUSIONS

This chapter has analysed the outstanding historic case of interrelated growth of output and productivity fed by the continuous introduction of endogenous technological innovations by Fiat during the years 1900–70.

To do this, a model of economies of growth based upon the interaction between localized technological change and growth has been elaborated. In this model, the growth of output leads to the valorization of localized competence and the endogenous introduction of technological change. The latter is now viewed as the driving force behind the relationship between growth and efficiency. When irreversibilities due to sunk costs and interrelatedness make extensive growth, based on increases in inputs, expensive, the accumulation of competence can lead to the sequential introduction of new and better technologies, characterized by high levels of complementarity, compatibility, and cumulability with existing production factors, which in turn increase efficiency and hence output.

The evidence of the Fiat case during the years 1900–70 confirms the relevance of this approach to understanding the interaction between technological change and growth. The empirical results of the Fiat case are important on three counts.

First, they suggest that the accumulation of intangible capital, consisting of experience and technological competence, and the eventual sequential introduction of cumulative technological changes, along a technological path shaped by the irreversibility of large portions of existing production factors, plays an important role.

Secondly, they confirm that the systematic internalization of the technological opportunities spilling over into the local environment, both within the technological district of Turin and in the technological cluster centred upon mechanical engineering, and the creation and implementation of technological complementarities among an array of specific products and processes in the mechanical engineering industries have played a major role in this growth process. This has been true since Fiat's foundation in a very fertile industrial and technological context.

Thirdly, a standard production function approach can be used in longitudinal analysis, but only when the shift in general efficiency, generated by the accumulation of competence and the resulting introduction of localized technological changes and creation of technological complementarities, is taken into account. The inclusion of a specific variable to capture the effects of technological change provides support to the hypothesis of dynamic increasing returns due to the generation of new technological knowledge and the introduction of technological innovations impinging upon localized 'bottom–up'

The Case of Fiat

processes of accumulation of competence and technological skills (Chandler, 1990, 1992).

Analysis of the rate and direction of endogenous technological innovations introduced by Fiat during the years 1900–70 highlights the important role of localized accumulation of technological competence, fed by efficiency wages and shaped by a cumulative sequence of systemic imbalances between product and process innovations applied to a widening variety of components for transportation equipment.

This case confirms the key role of technological complementarities for understanding the growth of firms. Technological complementarities are the result of the localized search for new products and processes in a technical space shaped by irreversible production factors. Technological complementarities supply the opportunity to take advantage of economies of scope and externalities. Economies of scope can be valorized by means of growth by diversification and vertical integration. Externalities can be valorized by means of localization within technological districts.

The empirical analysis has made it possible to substantiate the building process of organizational capabilities in the management of innovation activities. Fiat's technology management emerged gradually as a distinctive form of organizational capability, and it was based upon the identification of four distinct forms of technological knowledge—internal and external and tacit and codified—and the implementation of a dedicated style in managing the communication interfaces among each of them. The generation of localized technological change was made possible by the systematic absorption of (1) human capital and the implementation of skills which eventually brought about the accumulation of competence and technological knowledge, mainly built upon learning processes and reflective participation in the production process by a motivated workforce; (2) external knowledge available in the technological district of Turin; and (3) technological opportunities stemming from the widening applications of the technological cluster based upon mechanical engineering.

The systematic internalization of external knowledge, properly combined with internal accumulation of tacit knowledge based upon learning processes, made possible the introduction of localized technological change which increased the competitiveness of the company and hence output growth. This process has been further augmented by a dynamic macroeconomic context characterized by fast growth of revenue, high levels of income elasticity, and rapid diffusion of cars and other transportation equipment.

9

New Directions in the Corporate Production of Technological Knowledge

A significant shift in the economic understanding of the factors at play in the generation of technological knowledge has taken place over the last twenty years. For a long time, the corporation has been considered the main locus for the accumulation of technological knowledge and the object of relevant internal economies of scale and scope. A new approach to understanding and modelling the organization of the generation of technological knowledge is taking place: one where different forms of knowledge are identified and their interaction is appreciated.

New technological knowledge is now viewed as the collective outcome of an intentional and dedicated communication process finalized to enhance the interaction between tacit, codified, internal, as well as external knowledge. The external context in which internal research activities are conducted, such as technological districts and technological clusters where collective pools of technological knowledge are available, now receives greater attention.

Parallel and significant changes are taking place in the organization of the production of new technological knowledge, where technological communication and enhanced interaction among learning agents within collective innovation networks play a major role and seem to lead to the partial demise of the vertically integrated large corporation specializing in conducting internal research and development activities.

Since the late 1950s in the USA and the 1960s in Europe, intramural research and development activities, conducted within corporations and by corporations, have received a great attention. The notions of research and development activities and of innovation activities coincided for almost thirty years. Since the late 1980s, the attention of practitioners and economic analysis has broadened and, next to research and development activities, a wider range of innovation activities are now considered relevant. External economies in the production of knowledge as well as learning processes internal to firms are under the scrutiny of economic analysis.

This chapter explores this parallel evolution in the organization of the production of knowledge and in the economic understanding of the generation of technological knowledge. To do this, it is organized as follows: Section 1 recalls the simple elements of the generation of innovations in early manufacturing stages. Section 2 presents the basic ingredients of the R & D model. Section 3 presents the key features of the new model based upon the notion of localized technological knowledge. The main results are summarized in Section 4.

1. THE CRAFT MODEL

The craft organization of the generation of new knowledge was practised in Europe and the USA for a long period up until the first decades of the twentieth century. The craft organization highlighted the basic role of learning and tacit knowledge in the generation of new knowledge. Here, tacit knowledge based upon learning was in fact the basic input into the generation of new knowledge. Internal labour markets, apprenticeships, and competencies were considered the main prerequisites for the successful generation of new technologies. Small firms could easily compete with and contribute significantly to the rate of accumulation of new knowledge and introduction of new technologies. Natality of new entrepreneurial firms played a role as a vector of introduction of new technologies. Accumulation of knowledge on which new firms were based was made on the job by prospective entrepreneurs. Regional agglomeration was an important factor and the very notion of industrial districts emerged in the effort to appreciate the role of external economies in the generation of new technological knowledge. It can be argued that craft organization of the production of knowledge was appropriate for the rapid advances in the engineering industries: from machine tools to cars and general purpose and specialized machinery. In these activities, in fact, the scientific and

New Directions

generic content was relatively small, while the role of tacit knowledge based upon bottom–up learning processes prevailed (Reich, 1985; Noble, 1977).

2. THE R & D MODEL

The R & D model supplants the craft organization of the generation of new knowledge in the first part of the twentieth century. The new paradigm was based upon two key issues, well articulated in the pioneering contributions of Kenneth Arrow (1962a, 1969) and Alfred Chandler (1990, 1992). Technological knowledge was considered to be the outcome of a top–down process, where scientific discoveries and general laws, elaborated in universities and other research institutions, were the basic and preliminary ingredient. Technological knowledge was eventually generated as the outcome of a process of application and specification of the general scientific laws already available. The large corporation was viewed as the key institution in the generation of new technological knowledge. Internal research and development activities, funded from the profits of corporations and by public subsidies, were considered to be the core of the generation of new knowledge. Tight vertical integration of research and development activities within the corporation was the basic organizational design.

This model led to the sequential articulation of research and development activities in basic science, applied research and development activities, and the clear division of labour, with universities mainly in charge of the former and internal corporate laboratories for the latter.

The top–down model emerged mainly in the USA after the Second World War as a single and coherent institutional system. The basic arguments were (1) that a sequence existed between scientific and technological knowledge where science preceded technology; (2) that scientific knowledge had a public-good character, in terms of non-excludability, non-divisibility, and hence non-appropriability. In such conditions, an economic system could generate consistent levels of new technological knowledge only when it could provide sufficient incentive to agents to generate new knowledge (Mowery, 1983, 1989, 1995).

The model was articulated in seven main complementary aspects: (1) science was a public good and yet preceded technology and as such public subsidies to universities and to a lesser extent to private laboratories were necessary; (2) in order to increase appropriability, a strong intellectual property rights regime was also deemed necessary; (3) scientific communication among researchers was spontaneous and mainly secured by the circulation of postgraduates with Ph.D.s from universities to private laboratories; (4) the large corporation could fund R & D activities to implement generic scientific principles elaborated in universities and eventually generate technological changes. In this institutional context, large corporations had a strong competitive advantage because of: (5) important economies of scale and scope in conducting research activities and (6) their pre-existing monopolistic market power and a stable oligopolistic industrial structure where innovations, mainly product innovations, could be used, as competitive tools, and better appropriated within the framework of monopolistic competition. Finally, (7) R & D activities were mainly centralized within corporations: large multi-research laboratories were common practice. Laboratories were often located next to headquarters. Managerial responsibility for their conduct and funding was frequently assigned to the central staff at the headquarters. Plants and divisions were left with small R & D units performing mainly application tasks.

We contend that the R & D model was based upon much appreciative theorizing, elaborated in the 1950s and 1960s, of empirical evidence mainly provided by the chemical revolution of that time. These analyses and understandings of the generation of new technological knowledge were based upon the generalization and codification of stylized facts drawn from the actual historic trends traced by advances in chemistry and biology and in the accompanying technological changes resulting from the widespread application of chemistry to industrial production. In the chemical and biological sciences, scientific knowledge actually preceded technological advancement. The former was based upon actual discoveries and the latter was based upon technological testing and screening.

Enforcement of strong intellectual property rights was complementary to such a model and was considered an important institutional device to help prospective investors to channel appropriate levels of resources into the generation of new technological knowledge. Universities were the large common pool of scientific knowledge that firms and entrepreneurs could access in order

New Directions

to initiate their internal generation of technological knowledge and hence technological change (Machlup and Penrose, 1950; Machlup, 1962; Hirshleifer, 1971).

The pharmaceutical industry can be considered the 'stylized industrial organization and institutional set-up' of this model: universities primarily conduct basic research and provide scientific discoveries which are eventually translated into new products within the research laboratories of large corporations. Their size and efficiency provides the opportunities to take advantage of substantial increasing returns and hence barriers to entry which, in turn, increases the appropriability of new products which are, moreover, patented. Oligopolistic rivalry among firms in well-defined product markets determines the share of extra profits, part of which are used to fund further research and development activities (Mueller and Tilton, 1969; Scherer, 1984).

3. THE EMERGENCE OF THE LOCALIZED KNOWLEDGE MODEL

The demise of the Chandlerian mode of organization of knowledge production, based upon large corporations, and its partial substitution with networking innovation systems parallels the new theorizing about the localized knowledge model as opposed to the Arrovian top–down approach to understanding the generation of new technological knowledge.

Technological knowledge is increasingly viewed as the result of a bottom–up process of recombination of existing knowledge, both internal and external to each firm, accumulation of tacit competence, highly idiosyncratic, eventually blended with generic scientific knowledge. In the localized knowledge model the generation of technological knowledge is driven by the dynamic mixing of internal and tacit knowledge with external and codified knowledge (Ciborra, 1993; Gibbons et al., 1994).

The active search for technological complementarities, both synchronically and diachronically, that is, both with respect to the stock of tangible and intangible capital at each point in time and with respect to the knowledge-based activities of other firms, becomes the first and major axis for assessing and focusing the research strategy of firms.

In this context, the conditions for accessing external knowledge, within local and technological innovation systems, become key factors in the generation of new knowledge (Nelson, 1993; Lundvall and Johnson, 1994). Technological communication, however, is made difficult by the variety and complexity of details and applications in which new knowledge is embedded. The circulation of technological information appears more and more to be the result of intentional efforts rather than a spontaneous process. Firms actively build communication systems using technological cooperation with other firms and with universities, locating their research units within technological districts. Firms use trade-disembodied technology and rely on technological outsourcing. They stimulate employees to participate in the innovation process by activating the bottom–up process of internal technological communication. In fact, because of new awareness of the cumulability, compatibility, and complementarity of different kinds of technological knowledge, technological communication becomes more and more necessary for further advances to take place in the generation of technological knowledge and in the eventual introduction of further technological changes.

Industrial cooperation between manufacturing firms is more and more important in this context. Cooperative research procedures among independent firms, especially in the design of new complex products based upon a variety of components, are now common practice in the manufacturing industry and provide a variety of examples of forms of exchange and trade of both tacit and codified knowledge among teams of experts, each specializing in the design of complementary components. Discrimination in the conditions for accessing and using new knowledge produced within technological clubs is also becoming a major institutional device to increase reliability and effectiveness in cooperation. Partners agree upon the design of selective forms of access to the outcome of collective research. Such procedures are based upon differentiated time lags as well as partial access to the general outcome. In turn, participation in technological clubs is increasingly based upon differentiated forms of contribution (Arora and Gambardella, 1990, 1994).

Technological outsourcing is the third key element of the new model of knowledge production. Knowledge-intensive business services are now provided in the market-place and firms can rely on them for the procurement of specialized knowledge-intensive inputs which can be purchased rather than produced in-house. The interaction between manufacturing and service firms becomes a second key condition to achieving fast rates of generation of new technological knowledge and to the eventual introduction of technological innovations. Systematic technological outsourcing is also practised among innovative firms and takes the form of the trading of disembodied technological knowledge.

Recent empirical analyses show that the payments of the technological balance of payments of all the most advanced OECD countries have been increasing steadily since the late 1980s, both in absolute terms and with respect to domestic R & D expenditures, and now play a strong role in explaining rates of growth of total factor productivity (Antonelli, 2000c).

Firms sell and buy technological knowledge in a variety of forms such as trade in patents, licences, swapping of know-how, and technological assistance. Intellectual property rights are now mainly viewed as signalling and trading devices which help to localize the owners of specific forms of knowledge and make transactions easier (Ordover, 1991).

Fourthly, cooperation between universities and firms is more and more commonplace and firms rely on universities not only as a mean to access codified knowledge, germane to a variety of idiosyncratic uses, while tacit knowledge is elaborated internally, but as partners in the generation of new technological knowledge. The quality and access conditions to university research centres become a major factor in selecting the location of private research centres (Rosenberg and Nelson, 1992). Universities generate more and more knowledge which is directly relevant to the production of new technologies and firms often contribute to the advance of scientific knowledge. To a large extent, the distinction between academic research and business research no longer consists in the scientific content but in the incentive structure: whether it is aimed at feeding the publication process or can be kept secret or protected by intellectual property rights in the interest of the founder. At the same time, university budgets are increasingly funded by contracts awarded by firms within research programmes aimed at the generation of technological innovations which are directly relevant in the production process and in the marketing of new products (Geuna, 1999).

The traditional separation between academic research and business enterprises, research and development activities appears less and less appropriate and less and less efficient. The active involvement of universities and other academic institutions in the accumulation of new technological knowledge seems an important input for the eventual generalization of deductive understanding and hence the further implementation of new technological knowledge. At the same time, firms can take advantage of the skills and experience of academic researchers and academic researchers can have more direct and interactive access and active participation in the processes for generating applications upon which technological knowledge can be implemented.

Agglomeration of research activities within technological districts is also part of the new mode of knowledge production. Firms select the appropriate sites to locate their R & D laboratories globally, according to local endowments in terms of markets for skilled personnel, scientific and technological infrastructures, communication systems in place, opportunities for technological socialization with other complementary partners, co-localization of other advanced research centres, supply of advanced knowledge-intensive services. Regional specialization emerges as a powerful self-reinforcing mechanism for a few locations worldwide which are able to attract a growing number of research institutions.

In this context, learning plays a key role and without the intentional efforts of all the agents participating in the production process, at all levels of the hierarchical structure of the firm, it cannot take place. Learning is no longer considered as a free good or windfall. The stimulation and valorization of learning requires specific organizational procedures and an incentive structure. Only when such organization is in place, can firms actually learn and the positive effects be transformed into the accumulation of effective localized technological knowledge and the eventual introduction of new technological innovations.

Rates of generation of localized knowledge depend heavily on learning processes and their cumulativeness. This focuses attention on the role of employees in most advanced corporations, in terms of levels of involvement, active participation in the production process, and intentional efforts directed towards the accumulation of organizational and technological capital. Active participation and the contribution of the emotional and intellectual efforts of a qualified workforce in implementing learning processes make it possible to accumulate and better valorize tacit knowledge and experience, enabling the proper evaluation of the specific context of action, and enhancing the match between the availability of new codified knowledge and the experience of each firm. Rates of implementation of know-why, know-how, know-where, and know-when rest on the levels of participation by the skilled workforce in both production and decision-making.

This relationship seems to play a strong role, especially in knowledge-intensive business services where innovation often consists in the generation of customer-tailored specific and highly idiosyncratic solutions. Such innovative solutions can be elaborated only by involved and creative employees who are able to combine generic knowledge and hence high levels of human capital with a dedicated effort to actually understand the specific business conditions of customers.

Dynamic efficiency wages become especially relevant. Dynamic efficiency wages enhance loyalty and commitment and stimulate practitioners to develop informal relations and better collective work, sharing information and accelerating the emergence of tacit knowledge. Effective internal labour markets which favour the upgrading of competent employees within the firm are an important complementary tool to accelerate the rates of accumulation of experience and tacit knowledge. In fact, such markets keep competent labour within the firm and act as a powerful incentive to stimulate the participation of the workforce in learning processes. Dynamic efficiency wages increase learning rates, for they help to retain skilled manpower and extend the time period available to the firm to valorize the tacit knowledge acquired by each competent employee and blend it with the codified and the tacit knowledge of the other members of the organization. In sum, corporations are more and more aware that efficiency wages can significantly feed the accumulation of localized knowledge because they can stimulate the opportunity for internal mobility of motivated employees able to participate creatively in problem-solving activities, and activate the inductive processes of learning by doing, learning by using, and learning to interact with customers and in procurement.

In parallel with the new trends towards technological outsourcing and technological learning, major shifts are taking place in the organization of the generation of new knowledge within corporations.

Large centralized R & D laboratories are being replaced by internal networks of learning units. Smaller R & D laboratories have been created within divisions and in direct contact with and proximity to manufacturing plants. New forms of quasi-vertical integration are also emerging. Specialized research laboratories have been localized in other countries to benefit from local knowledge externalities and better command the flow of technological communication through the socialization of researchers within science parks and technological districts. Central R & D laboratories have been reorganized with coordination objectives in order to increase the general command of multi-site learning activities (Pavitt, 1987, 1991, 1999).

Systematic acquisition of high-tech start-ups has become common practice as a way to extend the stock of knowledge available. The traditional choice between make or buy is here extended to 'make, buy or take over'. Instead of purchasing external disembodied knowledge, large corporations, more and more, take over new high-tech companies which have already performed the crucial role of technological converters. They have, in fact, already done a number of essential tasks, such as: preliminary screening of new scientific and technological opportunities, technological testing, and the commercial assessment of new products and new processes. New start-ups have already acquired experience and competence in new interfaces between scientific advances and technological applications and have activated a variety of specific communication channels with dedicated customers and suppliers. The acquisition of high-tech start-ups is more and more a key tool to acquire external knowledge embodied in new organizations rather than in products.

Finally and most important, corporations are increasingly spinning off specialized units, especially in knowledge-intensive service activities, not only as a way to better use dedicated skills and manage excess capacity, but also to activate triangular flows of communication.

The spin-off of knowledge-intensive service units has many advantages from the managerial viewpoint. Spinning off units induces more active search for new customers beyond the borders of corporations and, in many instances, evidence has shown that the share of external demand quickly grows and becomes larger than captive, intra-corporate demand. The mixing of external demand with internal demand, however, has other and possibly stronger advantages in terms of the activation of triangular flows of technological communication. New spun-off units can better provide services and goods to holding companies' competitors and enter into user–producer interactions which full integration within the perimeter of the holding company would have prevented. User–producer interactions, in turn, activate learning and listening processes where the 'quasi-independent' specialized units become aware of their customers' specific problems, which are often similar to those of holding companies, and acquire the skill and competence to solve both sequentially.

As a result, the organization of knowledge production is moving away from the high levels of vertical integration that became the dominant mode of organization after the Second World War and is now directed towards the progressive unbundling of the production of knowledge.

Specialization based upon the notion of competence implemented with new institutional design is now becoming the leading organizational mode, engendered and implemented by the diffusion of electronic markets for technological knowledge. Firms have elaborated a variety of instruments to increase technological communication by means of new forms of connectivity and receptivity, enhancing organizational procedures such as:

1. the reorganization of R & D with internal multi-site research networks based upon a variety of small specialized and interactive R & D laboratories, under the control of divisional management, often located in regional innovation systems and in close proximity to manufacturing plants;
2. increased levels of technological outsourcing with specialized knowledge-intensive suppliers specializing in managing the complex interface between external tacit and codified knowledge and internal knowledge;
3. intensive participation in multi-disciplinary technological clubs;
4. the systematic search for external disembodied technological knowledge and its purchase from original innovators as an input for further recombination;
5. close interaction with universities and other public research centres for scientific and technological exchanges, permanent training, and outsourcing of specific research projects;

6. global selection of sites for the location of research and development laboratories within technological districts;
7. extensive use of dynamic efficiency wages in order to stimulate reflective employees to contribute to the innovative process;
8. systematic acquisition of high-tech start-ups;
9. spinning off of specialized knowledge-intensive units in order to activate triangular flows of technological communication;
10. enhanced interactions between marketing and research activities to capitalize upon better information on customer needs made available by online communication systems.

The diffusion of communication technologies has been an important factor in the spread of the new mode of technological knowledge production both among firms, between firms and consumers, and within corporations.

Technological communication among innovators plays an increasing role. Within technological clubs, laced together by online communication systems, firms can maximize the efficiency of the production of knowledge when economies of size matter, spreading the benefits to a large number of complementary users, yet keeping control over the conditions of appropriability implemented by *ex-ante* contractual agreements and enforced by reputation-building and signalling. In this context, applications of communication technologies, enforced by a rich array of contractual devices, are creating effective quasi-internal markets for the actual exchange and trade of technological information and technological knowledge. The opportunistic behaviour of partners can be sanctioned, not only by compromising the reputation of members as faithful or trustworthy, hence creating barriers to joining other clubs, but also and most important, by exclusion from the inter-corporate networks of communication and exchange on which technological cooperation is based, that is, mutual access to data banks, proprietary software, electronic archives, and such like. These cooperative structures are creating new opportunities for accumulating and exercising market power which is limited only by the proliferation of technological opportunities that allow rival clubs to pursue alternative technological strategies (Antonelli, 1999a; Antonelli et al., 2000).

Online communication systems make it easier for firms to identify the specific needs and routines of consumers and to better direct the generation of customized product innovations. This, in

New Directions

turn, induces firms to reorganize internally the flow of information and especially to strengthen communication channels between marketing, research, and communication activities.

The new directions of technological knowledge in the generation of communication technologies highlight the key role of technological convergence of a variety of scientific and technological disciplines and forms of knowledge as well as the application of a few general laws to a huge variety of specific contexts. The specific character of the accumulation of new knowledge in this area calls attention to the role of technological communication and recombination of incremental elaborations and specifications as a main source of additional knowledge. In many ways, communication technologies can be considered today to be the main vector of 'stylized facts' about present trends in the direction of technological knowledge and in the organization of the production of knowledge at large.

4. CONCLUSIONS

Understanding of the process of technological knowledge generation has undergone a major shift in the last twenty years. The model based upon the top–down process of application of generic, scientific principles into the variety of contingent conditions of operation of industrial firms is now being substituted by a bottom–up localized model of technological knowledge where the role of learning, socialization, and recombination, together with general scientific principles, is appreciated.

The R & D model assumed that technological knowledge had a strong public-good character because of high levels of nonexcludability, indivisibility, and non-appropriability of the generic scientific principles it rested upon. The large monopolistic and diversified corporations of the Chandlerian tradition were the 'best' institutions because of their *ex-ante* and *ex-post* monopolistic power and internal markets for the exchange of knowledge, which created incentives for the application of new scientific principles to economic activity.

Today, this model seems to be based upon the appreciative theorizing of historic evidence about the institutional and organizational set-up of a long period of economic and technological history dominated by the rapid advance of technological knowledge in the fields of chemistry and biology. It may now be obsolete.

Localized technological knowledge, as elaborated in specific contexts of application by well-defined clubs of players, has a much stronger collective character but only within specific contexts. By the same token, and because of the important role of cumulability, access to proprietary knowledge is more and more important as a condition for the generation of new knowledge: hence the growing need to examine whether existing structures of intellectual property protection or competition policy are appropriate for creating a diffusion-oriented learning economy. Participation in 'footloose' technological clubs and active localization in technological districts with open involvement in local flows of technological communication are more and more practised by corporations. Such strategies, as a matter of fact, are based upon the intentional reduction of intellectual property rights. This reduction takes place only if and when the conditions for reciprocal and possibly symmetrical access to technological communication is possible. Such strategies lead to the emergence of localized pools of collective knowledge (Ergas, 1987; Metcalfe, 1995a; Mazzoleni and Nelson, 1998).

Collective innovation, fed by localized interaction among an institutional variety of learning players within multi-disciplinary technological networks, seems to emerge as the new dominant mode of organization of the production of new knowledge. The understanding of laws governing horizontal and transversal circulation, dissemination, and recombination of technological knowledge is becoming much more important.

The borders of innovative corporations are becoming more and more porous and flexible. Exclusions and inclusions of business units are more and more frequent, with processes of diversification and specialization driven by the search for external knowledge.

10

Conclusions

The economics of innovation is facing the emergence of a new approach to understanding the microdynamics of technological change which parallels the diffusion both in the history and the sociology of science and technology of the social constructionism school (David, 1992, 1993, 1994; Bijker *et al.*, 1987; Smith and Marx, 1995; Latour, 1987; Misa, 1995). Social constructionism, as articulated by Latour (1987), contrasts with the deterministic model based upon technological trajectories and stresses the role of the collective undertakings of a myriad of innovators. The generation and introduction of technological innovations are now viewed as the result of complex alliances and compromises, based upon the valorization of weak knowledge indivisibilities and local externalities and hence potential complementarities among pieces of technological knowledge and technological innovations, among heterogeneous groups of agents. Agents are diverse because of the variety of competencies and localized bits of knowledge they build upon. The convergence of the efforts of a variety of innovators, each of which has a specific and yet complementary technological base, can lead to the successful generation of a new technology. In this approach, the distinction between innovation and diffusion is blurred and, by contrast, each adoption is viewed as the result of a complementary effort that makes a new technology useful and specifically reliable, increasing its scope of application. Adopters are no longer viewed as passive and reluctant prospective users, but rather as ingenious screeners that assess the scope for complementarity and cumulability of each new technology with their own specific needs and contexts of action. Profitability of adoption is the result of a process rather than a given fact. A technology diffuses when it applies to a variety of diverse conditions of use. The intrinsic heterogeneity of agents apply in fact not only to their own technological base but also to the product and factor markets in which they operate. The vintage structure of their fixed costs and both tangible and intangible capital can be portrayed as major factors of differentiation and identification of the specific context of action, both with respect to technological change and market strategy. New ideas can be implemented and incrementally enriched, so as to become profitable innovations eventually, when appropriate coalitions of heterogeneous firms are formed (Antonelli, 1995, 1999*a*).

A systemic approach to understanding the process of generation of new technological knowledge and introduction of new technologies seems relevant on many counts. System analysis is necessary because it makes it possible to appreciate the endogenous and pervasive role of synchronic and diachronic, product and process, internal and external technological complementarities. Prices are able to convey only a limited amount of the relevant information about the effects of complementarities on the static and dynamic behaviour of agents. In particular, prices are unable to convey all relevant information when firms are expected to be able to change their production functions and their technologies in out-of-equilibrium conditions. Innovations change the product and factor market conditions of firms. Their adjustment is constrained by the irreversibilities of their tangible and intangible capital stocks as well as their complementarities with other firms. The standard adjustment of prices to quantities is impeded and out-of-equilibrium conditions persist. In these circumstances, however, a creative reaction by firms may occur in terms of the introduction of new localized technologies when pools of collective knowledge are available and effective channels of technological communication are in place. Such localized technologies, introduced because of out-of-equilibrium conditions stemming from irreversibility, exhibit high levels of complementarity. Knowledge complementarities play a major role in this context.

The localized introduction of new technologies induced by irreversibilities leads to the emergence and implementation of technological systems. Technological systems are defined by a chain of technological complementarities that are the result of a localized search for new technologies around existing irreversible production factors. Technological complementarities can be synchronic and diachronic. Complementarities are synchronic when they characterize products and processes that are being introduced at the same

Conclusions 197

time. They are diachronic when they take place among different vintages of products and processes.

A system dynamics approach to understanding technological change makes it possible to overcome the limitations of the traditional demand-pull/technology-push macroeconomic analysis, as well as the deterministic notion of technological trajectories. It also makes it possible to root the analysis of technological analysis in a broader out-of-equilibrium context: technological knowledge is generated and technological change is introduced when firms cannot simply respond to changing conditions in their product and factor markets with adjustments of qualities to quantities. It provides, in fact, a broader context of analysis where the systemic features of technological knowledge are appreciated and combined with the industrial dynamics of firms able to generate new technological knowledge and to introduce technological innovations according to the specific conditions of the technological environment in which their market and technological strategies are embedded. The access conditions to collective knowledge available within technological clusters and technological districts play a key role. The eventual introduction of successful innovations, in turn, affects directly and indirectly the market conditions of all other firms and an ongoing process of technological and economic change is put in motion.

Evidence about the regional and technological concentration of innovation activities, an important result of the large empirical literature gathered in recent years, is consistent with the new understanding of the market construction of new technological systems. Innovation activities cluster in a few regions worldwide and in a few technologies. The increasing globalization of the international economy makes the regional concentration of innovation activities even more important: worldwide, firms concentrate their innovation activities in a few regions and in a few technologies, while distributing manufacturing and commercial activities in a much broader range of regions.

Evidence about the regional and technological concentration of innovative activities urges us to provide an interpretative framework. New growth theory acknowledges the role of learning and increasing returns in the production of technological knowledge, but assumes that they are at work systematically and automatically in advanced economic systems in an equilibrium context.

Evidence about concentration requires a broader and yet more specific framework.

The basic hypothesis put forward in this book is that increasing returns in the production of knowledge apply, albeit in a very specific and contextual set of complementary conditions. Such conditions can be found when a systemic approach to understanding the dynamics of technological knowledge and technological change is elaborated. The intrinsic indivisibility of technological knowledge puts a strain on the individualistic approach.

Increasing returns in the production of technological knowledge can take place within technological districts and technological clusters where qualified interactions among connected innovators make it possible to take advantage of the modular indivisibility and cumulativeness of technological knowledge.

To demonstrate this, the book has elaborated upon the notions of technological systems, technological complementarities, irreversibility, and collective knowledge. It has explored the role of learning, and technological communication, as key factors in the definition of the rate and direction of technological change within technological systems. Technological systems, such as technological districts and technological clusters, have been identified within a theoretical framework which values local externalities, irreversibility, and endogenous structural change.

Irreversibility matters and reduces the capability of firms to cope with the changing conditions of product and factor markets. When irreversibility matters, firms are induced to change the shape of their isoquants rather than moving along existing ones. Firms change their technology and modify their production conditions. Creative reactivity, or exaptation, as opposed to adaptation, has been one of the underlying themes of our analysis. Limitations to adaptations have been explored and the conditions for reactivity assessed.

Technological systems are the result of the chains of synchronic and diachronic, product and process, internal and external technological complementarities which characterize the localized technological innovations introduced by firms induced to innovate by relevant irreversibilities.

Local externalities, as opposed to global externalities, matter when the effort to actually assimilate external knowledge is accounted for and the role of proximity in technological communication is appreciated. Following the notion introduced by

Herbert Simon of near-decomposability of complex systems, knowledge can be thought to be articulated in subsystems characterized by high levels of viscosity and high-frequency dynamics of interactions. Knowledge complementarities matter mainly within modules and, as a consequence, technological externalities are available only locally, within technological districts and technological clusters, as the result of a complex web of communication channels and procedures which all need careful examination.

Regions are, at the same time, a source of major hysteretic constraints and innovation opportunity. Location is a major determinant of long-term rigidity and irreversibility as well as a context for technological communication, a source of external knowledge and learning opportunities: hence an important factor in assessing the rate and direction of technological change. Regions are a major factor in making technological change hysteretic. The interaction of the dynamics of localized technological changes, communication processes, and industrial dynamics explains the clustering of innovations in well-defined technological districts and technological clusters as well as the rate of introduction of technological changes.

This book provides an interpretative framework which identifies, qualifies, and specifies the very specific context and process conditions which make it possible for increasing returns to apply in the localized production of technological knowledge and technological change.

Technological knowledge is a systemic activity, rather than a good. Information is good while knowledge is an ongoing communication-intensive activity based upon recombination and learning, where individual agents are strongly interdependent both *ex ante* and *ex post*. The communication and recombination of internal and external knowledge play a key role in qualifying the basic characteristics of this activity. The generation of technological knowledge can exhibit increasing returns when and if this activity assumes a strong collective character as the result of specific innovation systems. Technological knowledge exhibits increasing returns within technological systems such as technological districts and technological clusters, when and where high levels of technological complementarities can be identified and the interactions are conducive to fostering technological communication among learning agents.

Secondly, it has been argued that learning and innovation, within technological districts and technological clusters, are characterized by disequilibrium conditions where firms need to cope with entropic change in their business environment and yet face production processes that are shaped by high levels of irreversibility of production factors. The accumulation of technological knowledge and the eventual introduction of technological change are then viewed as the result of the creative reaction of firms found in disequilibrium conditions. Creative reaction leads to the introduction of technological change when and if a conducive technological environment is available, one featured by high levels of technological communication and hence access to common pools of collective knowledge. Firms cluster in a few commons and in a few technological fields when and if common resources are not given and limited, but can be created and accrued with increasing returns.

Regions and local innovation systems, characterized by high levels of superfixed production factors and technological complementarities, business turbulence, and conducive conditions for technological communication, are likely to experience fast rates of introduction of technological changes. Location within well-defined regions becomes at the same time a factor of irreversibility and an inducement to the introduction of technological changes and a factor favouring the generation of technological knowledge in highly productive conditions. A local recursive hysteretic process is likely to take place in these circumstances. Much evidence, provided by industrial economics and economic history, about the increasing specialization of regions in the use of specific combinations of inputs finds a consistent interpretative framework in the model so far elaborated. The stochastic nature of local communication processes and their key role in long-term growth can play a major role in explaining discontinuities in such self-reinforcing mechanisms with sudden declines in local performance.

In this recursive context, the pervasive role of technological communication adds to the explanation of the dynamics of hysteretic and localized technological change. With high levels of mixed communication probabilities and hence innovation opportunities, firms' reactive and creative responses to all changes in their business environment favour the introduction of localized technological change. The better the access conditions to external technological knowledge which flow within communication systems, the higher are the chances for creative reaction to take place and hence the greater the opportunities for the accelerated introduction of localized technological changes (Freeman, 1991, 1997; Nelson, 1993).

Analysis of the interaction of the dynamics of localized technological changes and local communication processes provides important tools for understanding the clustering of innovations in well-defined regional spaces as well as the emergence of technological clusters characterized by the introduction of complementary innovations. This dynamics assumes all the characteristics of self-reinforcing processes fed by positive feedback and increasing returns. The detailed evidence now available in fact makes it possible to argue that increasing returns take place, not only within the fragile context of highly specific and necessary conditions, within technological districts and technological clusters, but also that positive feedback is at work and adds to the effects of static increasing returns with generalized 'Matthew effects'.

The notion of path dependence confirms here its heuristic power. The blending of the localized technological change approach and the broader issue of path dependence makes it possible to explain the role of both historic time and contextual factors. In a technological path, the probability of introduction of each new technology is contingent upon previous innovations as well as cumulated technological competence, but also on the necessary complementarity of other factors such as levels of irreversibility of the capital stock and access conditions to local knowledge externalities. The stochastic mix of complementary and necessary conditions affecting the transition from each step to the next along the path becomes key to understanding the actual sequence of events in assessing the rate and direction of technological change. Cumulative technological change can take place when firms are able to react to irreversibility traps with appropriate levels of technological creativity; when, because of effective communication systems, local externalities can turn into collective knowledge; when high levels of investment can help the introduction of new technologies; when industrial dynamics in product and input markets can induce localized technological changes which, in turn, affect the competitive conditions of firms; when stochastic processes help the creative interaction of complementary new localized knowledges and new localized technologies to form new effective technological systems; when the dynamics of positive feedback can actually implement the sequences of learning along technological paths, as well as the interactions between innovation and diffusion. Such a set of dynamic and systemic conditions has strong stochastic features and is available in finite conditions: the process is unlikely to continue indefinitely because of the exhaustion of the possible combinations. In these circumstances, the generation of new technological knowledge and the introduction of new technologies can be viewed as the cause and the consequence of punctuated economic growth and increasing returns. In this context, in fact, technological change can be viewed as a form of dynamic, systemic, stochastic, and finite increasing returns which leads to punctuated growth. Technological change takes place when a number of highly qualified necessary conditions apply: the successful introduction of technological change can be seen as the fragile result of a complex set of necessary and complementary conditions (David, 1975, 1985, 1997; Antonelli, 1999a; Arrow, 2000).

The approach so far elaborated seems useful in many ways. First, it provides a theory and an interpretative framework for understanding why local economic systems exposed to similar shocks, in terms of changes in factor and product markets, can react by introducing compatible technological changes which increase total factor productivity or assist in its eventual decline. We now have new elements for understanding the role of regions in assessing the persistent dynamic variety of firms within industries. The dynamics of local feedbacks between regions and firms helps explain why firms, located in different regions, react with different strategies and achieve different performances. The regional context of embedment and action of firms needs to be fully taken into account, both in terms of irreversibility and technological environment, when assessing industrial dynamics.

Agglomeration is not a sufficient condition for technological innovation to cluster and technological externalities to spill. Technological complementarities matter and are a necessary condition for technological communication to take place. A number of important communication channels are necessary and only their combination provides a conducive environment to foster the rate of accumulation of collective knowledge and eventual introduction of technological innovations.

A communication-based approach provides a theory for understanding the dynamics of agglomeration within technological

Conclusions 203

districts. The interplay between the collective character of technological knowledge and the characteristics of local communication processes is such that in technological districts, that is, regions with high levels of communication probability, the conditions for circulation and actual assimilation of technological information and the introduction of technological innovations reinforce each other with a self-reinforcing mechanism based upon the dynamics of positive feedback.

The dynamics of localized technological knowledge and communication processes can explain the interplay between the externality and the transaction costs approach. Low levels of transaction costs, namely communication costs, make it possible for local technological externalities to fully deploy their effects in terms of increasing returns and virtuous self-reinforcing mechanisms, where agglomeration, incremental accumulation of external stocks of technological knowledge, and economic growth reinforce each other. With high levels of communication costs, technological externalities exert only a limited positive effect and firms with given profitability conditions are much less inclined to fund research and development activities and eventually to introduce technological innovations. On the contrary, with effective multi-layer communication systems in place, technological externalities can effectively 'spill' into the air, albeit in a limited and contextual space, and recipients can take full advantage of the collective nature of technological knowledge, reducing its intrinsic dispersion and hence taking advantage of its complementarity and indivisibility.

Technological districts differ significantly from industrial districts. The latter are characterized mainly in terms of high levels of Marshallian externalities among small firms localized in regions with low levels of demographic concentration and small size of cities. The former, on the other hand, are qualified by the coexistence of large and small firms, a large multi-sectoral range of economic activities including both manufacturing and service industries, a strong metropolitan character, and a significant scientific and communication infrastructure. Technological districts can be considered to be the result of an evolutive process with respect to the industrial district. The accumulation of technological knowledge becomes central to the technological district as is the division of labour within manufacturing industry in the industrial district. Not all industrial districts can become technological districts. The growth of large firms seems to play a key role in such a transformation together with the local and spatially rooted supply of three important institutional factors: knowledge-intensive business services, universities, and financial services. The technological district can be identified as the spatial context where technological knowledge is both localized and collective. The opportunities provided by technological complementarities are implemented and translate into innovations and technological change by means of high-quality technological communication systems.

The notion of collective knowledge, in this context, has important implications. The dynamic properties of collective knowledge and technological communication make it possible to reconsider the tragedies of the commons. When collective knowledge and technological communication matter, access of each agent to a local common has a twin effect: on the one hand, it implies the use of common resources, but on the other hand, it helps increase the common pool.

The risks of overutilization—the so-called tragedy of the commons which according to standard analysis is the result of failure to assign property rights—are reduced when the positive effects of the entry and access of each agent are considered. The risks of overcircumscribing instead,—that is, the definition of property rights over indivisible resources—remain important and are likely to generate major dynamic inefficiencies preventing access to single bits of knowledge. A case for a tragedy of intellectual enclosures can emerge. When the collective character of knowledge within technological districts is implemented, technological innovations with strong elements of complementarity are introduced. Their sequential introduction, stimulated by the dynamics of positive feedback and positive externalities, is likely to constitute new technological clusters. Increasing returns in the production of knowledge can take place and here the tragedy of intellectual enclosures can become apparent. Intellectual enclosures can prevent firms from taking advantage of increasing returns in the production of knowledge and in the introduction of new technologies.

The rate of generation of new technological systems is strongly influenced by the localized complementarity and interdependence of the research agenda of a variety of learning organizations. Within technological districts, the accumulation of a wide common base of codified technological knowledge is germane to a myriad of specific

Conclusions 205

applications where the tacit knowledge accumulated by each agent in its specific context of action plays a key role. The relevance of such industrial and generic knowledge can be easily assimilated to the traditional economic analysis of 'commons'. The notion of collective technological knowledge as a basic intermediary good becomes more and more relevant in this context.

New technological knowledge is now viewed as the result of the implementation and recombination of long-lasting bits of pre-existing collective technological knowledge—yet dispersed among a myriad of possessors—which need to be mingled with the competence and tacit knowledge of know-how, know-why, know-when, and know-where to generate and introduce a new product and a new process. The outcome, rather than being a public good, is instead viewed as a collective process activity on account of its highly idiosyncratic characteristics which are specific to the localized scope of action of each agent.

Evidence of technological and structural change in communication technologies confirms the role of a system dynamics approach to understanding the process of generation of new technological knowledge and introduction of new technologies. The conduct and performances of firms are embedded in a web of constraints and opportunities caused by synchronic and diachronic complementarities and interdependencies that price analysis can appreciate only to a limited extent. Complementarities provide important opportunities to overcome the out-of-equilibrium conditions engendered by the introduction of new technologies by rivals, customers, and providers of intermediary inputs, in terms of the creation of new externalities, especially when adding to the common pools of collective knowledge. Complementarities cause major constraints when irreversibilities and sunk costs are taken into account because they limit rapid adjustment to new price–quantity combinations. Firms, however, are able to generate new technological knowledge and introduce new technologies. In this context, each innovation introduced by each firm has direct and indirect effects on both the conduct and performance of all the others. According to how well they cope with the new context, firms can reorganize their routines and may introduce new technologies feeding new waves of systemic change. Evidence of the close interplay between technological change and industrial dynamics in communication technologies is but an example of a general process of system dynamics where technological and economic change are fed by continuos feedback. The magnitude and relevance of the rates and directions of such changes depend upon the systemic conditions in which firms operate.

When such analysis is put in a broader context, it is also clear that innovation systems that are better able to activate the emergence of new technological districts and new technological clusters and to participate in technological convergence are likely to obtain larger levels of both comparative and absolute advantages and hence experience larger shares on international markets, both in the form of exports and in terms of multinational growth of domestic companies. In both cases, the positive effects in terms of employment and investment are enhanced.

In this context, the scope for intentional action is important. At firm level, three processes can be detected: (1) the search for new compensation schemes; (2) the elaboration of new technology strategies which enhance the capability to access external knowledge; (3) the active search for latent technological complementarities and their internalization by means of endogenous economies of scope.

The appreciation of learning leads firms to implement new forms of industrial relations and new compensation schemes: stock options and delayed remuneration based upon actual performance are becoming a common practice in most advanced countries. The goal is clearly to induce more active and direct participation by employees in the accumulation of new technological knowledge and the eventual introduction of technological changes. Dynamic efficiency wages are increasingly built into business practices.

The corporate organization of the production of knowledge is shifting away from the 'intramural' model based upon well-specified and self-contained research and development activities. A variety of tools are nowadays used by corporations to take advantage of external knowledge and minimize the tragedies of intellectual enclosures. Intentional participation in technological districts and technological clusters and business strategies characterized by flexible and porous borders appear to be practised increasingly by a growing number of corporations.

The evidence of the Fiat case seems instructive in this context. The systematic internalization of external technological knowledge, available both within the technological district of Turin and the technological cluster centred upon mechanical engineering, and the valorization of internal skills have been a driving force behind Fiat's accumulation of internal knowledge and technological capability. Here the growth of a large corporation seems, also, to be the result of specific competencies and managerial routines which have been able to keep open a variety of communication channels within the 'walls' of the company and with the external environment. To a large extent, the evidence of the Fiat case suggests that the Chandlerian model, based upon intramural research and development activities, has been much less relevant in the European context.

Corporations and districts appear as distinctive modes of organizing and coordinating the production of goods as well as of accumulating knowledge and introducing new technologies. Complex relations take place between corporations and districts. Corporations substitute for districts, such as in the Fiat case, especially when the dynamics of the technological system declines. In many circumstances, however, corporations and districts complement each other and large corporations concentrate their knowledge-intensive activities in technological districts next to small firms. This takes place especially when a new technological system is emerging. Finally, the decline and possible collapse of large corporations may lead to the re-emergence of technological districts in new areas of activity and technological specialization.

The model of a localized search for new complementary technologies, stimulated by the entropy of products and factors markets, has proved useful for integrating the theory of the firm in a post-Chandlerian perspective. Economies of scope, in this context appear as the result of the localized search for new technologies and hence the endogenous outcome of the localized capability to react creatively to a changing market when irreversibility limits sheer adaptation.

Identification of latent technological complementarities and their active implementation become major instruments for aligning the technological strategies of firms. Economies of scope are the endogenous result of the intentional strategy of firms able to take advantage of modules of technological complementarities and direct their innovation activities so as to make possible new productive uses of the portions of existing and irreversible stocks of tangible and intangible capital, including stocks of localized technological knowledge.

The dynamics of the gales of localized technological changes, within technological districts and technological clusters, stresses the role of the intentional search for complementarity and receptivity enhancing strategies by firms. Receptivity enhancing becomes the target of dedicated strategies at company level. Receptivity and connectivity can be intentionally implemented by firms which are more aware of the role of technological communication to better acquire and make use of external knowledge generated by other firms, reducing communication lags and the not-invented-here syndrome. The intentional search for external knowledge in a context of reciprocity and shared intellectual property becomes the axis for an active technological and business strategy by firms that are aware of the systemic characteristics of localized technological change.

At the aggregate level, the implications for industrial and innovation policies are far-reaching.

The first implication arises from analysis of the effects of dynamic efficiency wages on labour markets. The segmentation of workers according to their role in the accumulation of technological knowledge within companies can become a serious matter on many counts. A major divide can take place, between firms and within firms, across categories, skills, vintages, contents of codified human capital, in terms of wages and duration of labour spells. A strong trend towards a new dualism may emerge between tenured and well-paid reflective and responsible workers who are actively engaged in jobs with high learning potential, that is, jobs where the scope for learning is acknowledged and appreciated, and 'flexible' low-paid passive workers occupied in jobs with little prospect for actual learning to take place. Active training policies, with special attention to permanent training, can become important both to reduce the negative effects of the new dualism and to enhance the scope for learning at the system level.

A second and important issue concerns the role of innovation policy itself. In the Arrovian tradition, innovation policy is necessary to limit the negative consequences of market failures associated with the public-good character of technological knowledge. Public subsidies and intellectual property rights are necessary to reduce the damage of the market system in terms of underallocation of resources to inventive activity. When the new growth theory perspective is assumed, together with the qualifications and limitations we have introduced, innovation policy is no longer required as a

Conclusions 209

remedy to market failure, but becomes a tool permitting increasing returns to blossom and display the full strength of their effects. A proper combination of regional and innovation policy interventions can help reinforce the institutional and economic conditions which make it possible for localized technological knowledge to become a collective good and hence to give way to increasing returns and faster rates of introduction of technological change. Innovation policy becomes a tool for a market-enhancing strategy as opposed to a market-failure remedy. This calls for a context-based, as opposed to a context-free, innovation policy.

Identification of technological complementarities and appreciation of the factors governing technological communication and the internalization of local technological externalities among firms which are involved in complementary innovation activities provides guidance to elaborate and implement a possible strategy for public intervention. On these bases, a context-based innovation policy can be elaborated, directed towards realizing the advantages stemming from increasing returns in the production of technological knowledge and hence to an increase in the productivity of resources invested in innovation activities. More specifically, public subsidies aimed at enhancing technological communication in terms of better trade conditions for disembodied technological knowledge, local supply of knowledge-intensive business services, technological cooperation both among firms and between firms and universities, accelerated licensing of patents and know-how, can offer firms the opportunity to internalize the spillover of localized technological knowledge and take better advantage of available external knowledge with the active participation of both parties in the trade: vendors and customers. Enhanced rates of introduction of technological changes and faster rates of increase of total factor productivity may be obtained with the implementation of local communication processes.

When the collective components of technological knowledge are identified, together with the important role of external knowledge as an essential intermediary input, the characteristics of local innovation systems and specifically of technological districts, that is, regional clustering of learning firms, working in complementary technologies, in terms of the quality and effectiveness of all communication networks in which technological information is shared and is effectively transmitted from one firm to another, become relevant and warrant greater attention. Appreciation of the factors governing technological communication and effective internalization of local technological externalities among firms which are involved in complementary innovation activities becomes a possible strategy for public intervention, in that such activities will lead to an increase in the productivity of resources invested in innovation activities.

High levels of regional concentration of innovation capabilities within technological districts, as identified in this work, play a major role in fostering the rate of introduction of technological change at the country level. Such 'spontaneous' concentration of factors takes place, within a country, in a limited number of regions. Its 'artificial' reproduction seems a complicated task and can be achieved only in the long term. This leads us to wonder whether the classical debate about the Schumpeterian trade-off between static and dynamic efficiency might apply at a regional level as well. A large body of empirical evidence, gathered in industrial economics, suggests that while low levels of concentration are a condition for static efficiency to be achieved, oligopolistic rivalry among a limited number of large firms is more conducive to dynamic efficiency in terms of faster rates of introduction of innovations. In this context, it seems likely that concentration in a few technological districts of scientific and academic infrastructures and public subsidies might be considered an appropriate choice in order to sustain the rate of technological change at the aggregate level.

The identification of both actual and eventual technological districts clearly becomes an important issue. Regions characterized by: (1) actual and eventual coherence in terms of levels of technological complementarities among the knowledge base of the firms; (2) high levels of knowledge-intensive activities; (3) high levels of density of knowledge-intensive activities and (4) high-quality systems of technological communication, are likely to exhibit high levels of efficiency in the generation of new technological knowledge.

In a similar way, the identification of technological clusters that are especially relevant to national economic systems and where national players can play a role becomes an important condition for focusing on national innovation policy. A selective innovation policy is called for with respect to technological clusters. First, national innovation policy should focus on the segments of the industrial structure that are likely to constitute a technological cluster and take advantage of its system dynamics based upon increasing returns and positive feedbacks. Secondly, a limited number of such potential technological clusters can be assisted and their identification should be based upon: (1) levels of actual scientific and technological opportunities; (2) market opportunities for national firms; (3) externalities in downstream and upstream sectors which they are likely to generate, and most important, (4) the actual complementarities of firms with respect to both their stocks of localized technological knowledge and their current strategies of technological innovation.

A selective innovation policy which combines elements of industrial and regional policy and addresses the limited number of regions and technological clusters where increasing returns are likely to stem from collective knowledge can be elaborated on this basis.

Such selection can take place at two levels. From an aggregate viewpoint, it suggests that the concentration of public resources, available to sustain innovation activities, in a limited number of technological clusters and districts is appropriate and more useful than an even distribution across regions and technologies. From a more disaggregated viewpoint, at the regional level, a selective innovation policy can be implemented, specializing the focus of local innovation activities on those technological fields where stronger elements of irreversibility, complementarity, and accumulated local stocks of knowledge appear to be available.

The production of technological knowledge is characterized by increasing returns, but only at the system level, where technological complementarities matter, and within technological districts and technological clusters, when firms are able to cope with the disequilibrium conditions engendered by the irreversibility trap and are induced to valorize their learning procedures. It seems clear that economic systems able to aggregate and structure the economic, institutional and behavioural conditions that favour the emergence of technological districts and which specialize in the activities that impinge upon new technological clusters can benefit in terms of higher levels of total factor productivity, and hence lower costs, barriers to entry, and long-lasting quasi-rents.

References

Abernathy, W. J. (1978), *The Productivity Dilemma*, Baltimore: Johns Hopkins University Press.
Adler, P. (ed.) (1992), *Technology and the Future of Work*, Oxford: Oxford University Press.
—— and Winograd, T. (eds.) (1992), *Usability: Turning Technology into Tools*, Oxford: Oxford University Press.
Aghion, P., and Howitt, P. (1998), *Endogenous Growth Theories*, Cambridge, Mass.: MIT Press.
Ahmad, S. (1966), 'On the theory of induced invention', *Economic Journal*, 76: 344–57.
Akerlof, G. A. (1982), 'Labor contracts as partial gift exchange', *Quarterly Journal of Economics*, 97: 543–69.
—— and Yellen, J. L. (1986), *Efficiency Wages Models and the Labor Market*, Cambridge: Cambridge University Press.
Albin, P. S. (1998), *Barriers and Bounds to Rationality*, Princeton: Princeton University Press.
Allen, R. C. (1983), 'Collective invention', *Journal of Economic Behavior and Organization*, 4: 1–24.
Amendola, M., and Bruno, S. (1990), 'The behaviour of the innovative firm: relations to the environment', *Research Policy*, 19: 419–34.
—— and Gaffard, J. L. (1988), *The Innovative Choice: An Economic Analysis of the Dynamics of Technology*, Oxford: Blackwell.
—— —— and Musso, P. (1999), 'Competition, innovation, and increasing returns', *Economics of Innovation and New Technology*, 9: 149–81.
Anselin, L., and Varga, A. (1997), 'Local geographical spillovers between university research and high technology innovation', *Journal of Urban Studies*, 42: 422–48.
Antonelli, C. (1986a), *L'Attività Innovativa in un Distretto Tecnologico*, Turin: Edizioni della Fondazione Agnelli.
—— (1986b), 'Technological districts and regional innovation capacity', *Revue d' Économie Régionale et Urbaine*, 5: 695–705.
—— (1987), 'The determinants of the distribution of innovative activity in a metropolitan area: the case of Turin', *Regional Studies*, 21: 85–94.
—— (1990), 'Induced adoption and externalities in the regional diffusion of new information technology', *Regional Studies*, 24: 31–40.
—— (1993a), 'Externalities and complementarities in telecommunications dynamics', *International Journal of Industrial Organization*, 11: 437–48.

—— (1993b), 'The dynamics of technological interrelatedness: the case of information and communication technologies', in D. Foray and C. Freeman (eds.), *Technology and the Wealth of Nations*, London: Francis Pinter.

—— (1994), 'Technological districts, localized spillovers, and productivity growth: the Italian evidence on technological externalities in the core regions', *International Review of Applied Economics*, 8: 31–45.

—— (1995), *The Economics of Localized Technological Change and Industrial Dynamics*, Boston: Kluwer Academic Publishers.

—— (1999a), *The Microdynamics of Technological Change*, London: Routledge.

—— (1999b), 'The dynamics of technological systems: the case of new communication technologies', *Review of Economic Conditions of Italy*, 50: 245–86.

—— (2000a), 'Collective knowledge, communication, and innovation: the evidence of technological districts', *Regional Studies*, 34: 535–47.

—— (2000b), 'Restructuring and innovation in long-term regional change', in Clark *et al.* (2000).

—— (2000c), 'Recombination and the production of technological knowledge: some international evidence', in Metcalfe and Miles (2000).

—— (forthcoming a), 'Innovation in advanced telecommunications networks', in G. Madden and S. Savage (eds.), *The International Handbook on Telecommunications Economics*, Cheltenham: Edward Elgar.

—— (forthcoming b), 'Collective knowledge and learning to communicate', in M. Feldman and N. Massard (eds.), *Knowledge Spillovers and the Geography of Innovation: Institutions and Systems of Innovation*, Dordrecht: Kluwer Academic Publishers.

—— Geuna, A., and Steinmueller, E. W. (2000), 'Information and communication technologies and the production, distribution and use of knowledge', *International Journal of Technology Management*, 20: 75–94.

Aoki, M. (1988), *Information, Incentives, and Bargaining in the Japanese Economy*, Cambridge: Cambridge University Press.

Archibugi, D., and Michie, J. (eds.) (1997), *Technology, Globalisation, and Economic Performance*, Cambridge: Cambridge University Press.

—— —— (eds.) (1998), *Trade Growth and Technical Change*, Cambridge: Cambridge University Press.

—— and Pianta, M. (1992), *The Technological Specialization of Advanced Countries*, Boston: Kluwer Academic Publishers.

Arora, A. (1995), 'Licencing tacit knowledge: intellectual property rights and the market for know-how', *Economics of Innovation and New Technology*, 4: 41–60.

Arora, A. (1997), 'Patents licensing and market structure in the chemical industry', *Research Policy*, 26: 391–403.

—— and Gambardella, A. (1990), 'Internal knowledge and external linkages: theoretical issues and an application to biotechnology', *Journal of Industrial Economics*, 37: 361–79.

—— —— (1994), 'The changing technology of technological change: general and abstract knowledge and the division of innovative labour', *Research Policy*, 23: 523–32.

Arrow, K. J. (1962a), 'The economic implications of learning by doing', *Review of Economic Studies*, 29: 155–73.

—— (1962b), 'Economic welfare and the allocation of resources for invention', in R. R. Nelson (ed.), *The Rate and Direction of Inventive Activity: Economic and Social Factors*, Princeton: Princeton University Press for NBER.

—— (1969), 'Classificatory notes on the production and transmission of technical knowledge', *American Economic Review*, 59: 29–35.

—— (1974), *The Limits of Organization*, New York: W. W. Norton.

—— (1996), 'Technical information and industrial structure', *Industrial and Corporate Change*, 5: 645–52.

—— (2000), 'Increasing returns: historiographic issues and path dependence', *European Journal of History of Economic Thought*, 7: 171–80.

Arthur, B. (1989), 'Competing technologies, increasing returns, and lock-in by small historical events', *Economic Journal*, 99: 116–31.

—— (1994), *Increasing Returns and Path-Dependence in the Economy*, Ann Arbor: The University of Michigan Press.

Atkinson, A. B., and Stiglitz, J. E. (1969), 'A new view of technological change', *Economic Journal*, 79: 573–8.

—— —— (1980), *Lectures on Public Economics*, Maidenhead: McGraw-Hill.

Audretsch, D. B., and Feldman, M. (1996), 'Spillovers and the geography of innovation and production', *American Economic Review*, 86: 630–40.

—— and Stephan, P. E. (1996), 'Company-scientist locational links: the case of biotechnology', *American Economic Review*, 86: 641–52.

Bairati, P. (1983), *Valletta*, Turin: UTET.

Bania, N., Eberts, R. W., and Fogarty, M. S. (1993), 'Universities and the startup of new companies: can we generalize from Route 128 and Silicon valley?' *Review of Economics and Statistics*, 75: 761–6.

Baptista, R. (2000), 'Does innovation diffuse faster within geographical clusters', *International Journal of Industrial Organization*, 18: 515–35.

—— and Swann, P. (1998), 'Do firms in cluster innovate more', *Research Policy*, 27: 527–42.

Bassignana, P. L. (ed.) (2000), *Dante Giacosa: Il Mestiere di Progettista: Antologia degli Scritti*, Turin: Archivio Storico Fiat, Archivio Storico AMMA.

Becattini, G. (1979), 'Dal "settore" industriale al "distretto" industriale: Alcune considerazioni sull'unità di indagine dell'economia industriale', *Rivista di Economia e Politica Industriale*, 2: 7–21.

—— (ed.) (1987), *Mercato e Forze Locali: Il Distretto Industriale*, Bologna: Il Mulino.

—— (ed.) (1989), *Modelli Locali di Sviluppo*, Bologna: Il Mulino.

Bellandi, M. (1989), 'Capacità innovativa diffusa e sistemi locali di imprese', in Becattini (1989).

—— and Russo M. (eds.) (1994), *Distretti Industriali e Cambiamento Economico Locale*, Turin: Rosenberg & Sellier.

Beniger, J. R. (1986), *The Control Revolution: Technological and Economic Origins of the Information Society*, Cambridge, Mass.: Harvard University Press.

Bernanke, B. S. (1983), 'Irreversibility, uncertainty, and cyclical investment', *Quarterly Journal of Economics*, 97: 85–106.

Bessant, J., and Rush, H. (1995), 'Building bridges for innovation: the role of consultants in technology transfer', *Research Policy*, 24: 97–114.

Bijker, W. E., Hughes, T. P., and Pinch, T. (eds.) (1987), *The Social Construction of Technological Systems*, Cambridge, Mass.: MIT Press.

Boyer, R., Chavance, B., and Godard, O. (eds.) (1991), *Les Figures de l'Irréversibilité en Économie*, Paris: Éditions de l'École des Hautes Études en Sciences Sociales.

Bresnahan, T., and Trajtenberg, M. (1995), 'General purpose technologies: engines of growth', *Journal of Econometrics*, 65: 83–108.

Brusco, S. (1992), 'The Emilian model: productive decentralization and social integration', *Cambridge Journal of Economics*, 6: 167–80.

Calvo, G. A., and Wellisz, S. (1979), 'Hierarchy ability and income distribution', *Journal of Political Economy*, 87: 991–1010.

Camagni, R. (ed.) (1991), *Innovation Networks*, London: Belhaven Press.

—— (1999), *Innovation Networks: Spatial Perspectives*, New York: John Wiley & Sons.

Cantwell, J. A. (1989), *Technological Innovation and Multinational Corporations*, Oxford: Basil Blackwell.

Carlsson, B. (ed.) (1995), *Technological Systems and Economic Performance: The Case of Factory Automation*, Boston: Kluwer Academic Publishers.

—— (1998), 'On and off the beaten path: the evolution of four technological systems in Sweden', *International Journal of Industrial Organization*, 15: 775–800.

—— and Eliasson, G. (1994), 'The nature and importance of economic competence', *Industrial and Corporate Change*, 3: 687–712.

Carlsson, B., and Stankiewitz, R. (1991), 'On the nature, function, and composition of technological systems', *Journal of Evolutionary Economics*, 1: 93–118.

Carter, A. (1989), 'Know-how trading as economic exchange', *Research Policy*, 18: 155–63.

Castells, M. (1989), *The Informational City: Information Technology, Economic Restructuring, and the Urban-Regional Process*, Oxford: Blackwell.

Castronovo, V. (1971), *Giovanni Agnelli*, Turin: UTET.

Chandler, A. D. (1990), *Scale and Scope: The Dynamics of Industrial Capitalism*, Cambridge, Mass.: Harvard University Press.

—— (1992), 'Organizational capabilities and the economic history of industrial enterprise', *Journal of Economic Perspectives*, 6: 79–100.

Ciborra, C. (1993), *Teams, Markets, and Systems: Business Innovation and Information Technology*, Cambridge: Cambridge University Press.

Clark, G. L., Feldman, M., and Gertler, M. S. (eds.) (2000), *The Oxford Handbook of Economic Geography*, Oxford: Oxford University Press.

Clark, K., and Fujimoto, T. (1991), *Product Development Performance: Strategy Organization and Management in the World Auto Industry*, Boston: Harvard Business School Press.

Clarysse, B., Debackere, K., and Van Dierdonck, R. (1995), 'Research networks and organizational mobility in an emerging technological field: the case of biotechnology', *Economics of Innovation and New Technology*, 4: 77–96.

Cohen, W. M., and Levinthal, D. A. (1989), 'Innovation and learning: the two faces of R&D', *Economic Journal*, 99: 569–96.

—— —— (1990), 'Absorptive capacity: a new perspective on learning and innovation', *Administrative Science Quarterly*, 35: 128–52.

Coltelletti, L. (1991), 'L'Attività brevettuale della Fiat dagli inizi al 1930', in *Progetto Archivio Storico*, Milan: Fabbri.

Cowan, R., and Foray, D. (1997), 'The economics of codification and the diffusion of knowledge', *Industrial and Corporate Change*, 20: 595–622.

Dasgupta, P., and Stiglitz, Joe E. (1980), 'Industrial structure and the nature of innovative activity', *Economic Journal*, 90: 266–93.

—— and Stoneman, P. (eds.) (1987), *Economic Policy and Technological Performance*, Cambridge: Cambridge University Press.

David, P. A. (1975), *Technical Choice, Innovation, and Economic Growth*, Cambridge: Cambridge University Press.

—— (1985), 'Clio and the economics of QWERTY', *American Economic Review*, 75: 332–7.

—— (1987), 'Some new standards for the economics of standardization in the information age', in Dasgupta and Stoneman (1987).

—— (1992), 'Heroes, herds, and hysteresis in technological history', *Industrial and Corporate Change*, 1: 129–79.

—— (1993), 'Knowledge property and the system dynamics of technological change', *Proceedings of the World Bank Annual Conference on Development Economics*, Washington: World Bank.

—— (1994), 'Positive feed-backs and research productivity in science: reopening another black box', in O. Granstrand (ed.), *Economics and Technology*, Amsterdam: Elsevier.

—— (1997), *Path Dependence and the Quest for Historical Economics: One more Chorus of the Ballad of QWERTY*, University of Oxford Discussion Papers in Economic and Social History, no. 20.

—— (1998), 'Communication norms and the collective cognitive performance of "Invisible Colleges"', in G. B. Navaretti, P. Dasgupta, K. G. Maler, and D. Siniscako (eds.), *Creation and the Transfer of Knowledge: Institutions and Incentives*, Berlin, Heidelberg, and New York: Springer-Verlag.

—— and Foray, D. (1994), 'The economics of EDI standards diffusion', in G. Pogorel (ed.), *Global Telecommunications Strategies and Technological Changes*, Amsterdam: North-Holland.

—— —— and Dalle, J. M. (1998), 'Marshallian externalities and the emergence and spatial stability of technological enclaves', *Economics of Innovation and New Technology*, 6: 147–82.

—— and Steinmueller, E. (1994), 'Economics of compatibility standards and competition in telecommunication networks', *Information Economics and Policy*, 6: 217–42.

Dei Ottati, G. (1996), 'Trust interlinking transactions and credit in industrial districts', *Cambridge Journal of Economics*, 18: 529–46.

Dixit, A. (1992), 'Investment and hysteresis', *Journal of Economic Perspectives*, 6: 107–32.

Dorfman, N. (1983), 'Route 128: the development of a regional high-technology economy', *Research Policy*, 12: 299–316.

Downie, J. (1958), *The Competitive Process*, London: Gerard Duckworth.

Edquist, C. (ed.) (1997), *Systems of Innovation: Technologies, Institutions, and Organizations*, London: Pinter.

Engelbrecht, H. J. (1998), 'A communication perspective on the international information and knowledge system', *Information Economics and Policy*, 10: 359–67.

Ergas, H. (1987), 'Does technology policy matter?' in B. R. Guile and H. Brooks (eds.), *Technology and Global Industry: Companies and Nations in the World Economy*, Washington: National Academy Press.

Farrell, M. J. (1957), 'The measurement of productive efficiency', *Journal of the Royal Statistical Society*, Series A, 120: 253–81.

Fauri, F. (1996), 'The role of Fiat in the development of the Italian car industry in the 1950s', *Business History Review*, 70: 167–206.

Feldman, M. P. (1993), 'An examination of the geography of innovation', *Industrial and Corporate Change*, 2: 447–67.

—— (1994), *The Geography of Innovation*, Dordrecht: Kluwer Academic Publishers.

—— (1999), 'The new economics of innovation spillovers and agglomeration: a review of empirical studies', *Economics of Innovation and New Technology*, 8: 5–26.

—— and Audretsch, D. B. (1999), 'Innovation in cities: science-based diversity, specialization, and localized competition', *European Economic Review*, 43: 409–30.

Finch, J. H. (2000), 'Is post-Marshallian economics an evolutionary research tradition?', *European Journal of History of Economic Thought*, 7: 377–406.

Foray, D. (1991), 'The secrets of industry are in the air: industrial cooperation and the organizational dynamics of the innovative firm', *Research Policy*, 20: 393–405.

Fransmann, M. (1994a), 'AT&T, BT and NTT: a comparison of vision, strategy, and competence', *Telecommunications Policy*, 18: 137–53.

—— (1994b), 'AT&T, BT and NTT: the role of R&D', *Telecommunications Policy*, 18: 295–305.

—— (1995), *Japan's Computer and Communications Industry*, Oxford: Oxford University Press.

—— (1999), *Visions of Innovation*, Oxford: Oxford University Press.

Freeman, C. (1991), 'Networks of innovators: a synthesis of research issues', *Research Policy*, 20: 499–514.

—— (1997), 'The national system of innovation in historical perspective', in Archibugi and Michie (1997).

—— and Soete, L. (1997), *The Economics of Industrial Innovation*, 3rd edn., London: Pinter.

Gallouj, F., and Weinstein, O. (1997), 'Innovation in services', *Research Policy*, 26: 537–56.

Geroski, P. A. (1998), 'An applied econometrician's view of large company performance', *Review of Industrial Organization*, 13: 271–93.

Geuna, A. (1999), *The Economics of Knowledge Production*, Cheltenham: Elgar.

Gibbons, M., Limoges, C., Nowotny, H., Schwarzman, S., Scott, P., and Trow, M. (1994), *The New Production of Knowledge: The Dynamics of Research in Contemporary Societies*, London: Sage Publications.

References

Griliches, Z. (1979), 'Issues in assessing the contribution of research and development to productivity growth', *Bell Journal of Economics*, 10: 92–116.
—— (ed.) (1984), *R&D Patents and Productivity*, Chicago: NBER.
—— (1986), 'Productivity, R&D, and basic research at the firm level in the 1970s', *American Economic Review*, 76: 141–54.
—— (1992), 'The search for R&D spillover', *Scandinavian Journal of Economics*, 94: Supplement, 29–47.
Gould, S. J. (1991), 'Exaptation: a crucial tool for an evolutionary psychology', *Journal of Social Issues*, 47: 43–65.
—— and Vrba, E. S. (1982), 'Exaptation: a missing term in the science of form', *Paleobiology*, 8: 4–15.
Hagerdoorn, J. (1995), 'Strategic technological partnership during the 1980s: trends, networks, and corporate patterns in non-core technologies', *Research Policy*, 24: 207–31.
Harrison, B. (1992), 'Industrial districts: old wine in new bottles', *Regional Studies*, 26: 469–83.
—— Kelley, M. R., and Gant, J. (1996), 'Innovative behavior and local milieu: exploring the intersection of agglomeration firm effects and technological change', *Economic Geography*, 72: 233–50.
Hashimoto, M. (1981), 'Firm specific human capital as a shared investment', *American Economic Review*, 71: 475–82.
Hayek, F. A. (1945), 'The use of knowledge in society', *American Economic Review*, 35: 519–30.
Helpman, E. (ed.) (1998), *General Purpose Technologies and Economic Growth*, Cambridge, Mass.: MIT Press.
Henderson, R., and Clark, K. B. (1990), 'Architectural innovation: the reconfiguration of existing product technologies and the failure of established firms', *Administrative Sciences Quarterly*, 35: 9–30.
Henry, C. (1974), 'Investment decisions under uncertainty: the "irreversibility effect"', *American Economic Review*, 64: 1006–12.
Hicks, J. (1932), *The Theory of Wages*, London: Macmillan.
Hirschleifer, J. (1971), 'The private and social value of information and the reward to inventive activity', *American Economic Review*, 61: 561–74.
Honig-Haftel, S., and Martin, L. (1993), 'The effectiveness of reward systems on innovative output: an empirical analysis', *Small Business Economics*, 5: 261–9.
Howells, J. (1990), 'The location and organization of research and development: new horizons', *Research Policy*, 19: 133–46.
Hughes, T. P. (1984), *Networks of Power: Electrification in Western Society, 1880–1930*, Baltimore: Johns Hopkins University Press.

Ichiniowski, C., Shaw, K., and Prennushi, G. (1997), 'The effects of human resources management practices on productivity: a study of steel finishing lines', *American Economic Review*, 87: 291–313.
Jaffe, A. B., Trajtenberg, M., and Henderson, R. (1993), 'Geographic localization and knowledge spillovers as evidenced by patent citations', *Quarterly Journal of Economics*, 108: 577–98.
Kaldor, N. (1957), 'A model of economic growth', *Economic Journal*, 67: 591–624.
—— and Mirrlees, J. (1962), 'A new model of economic growth', *Review of Economic Studies*, 29: 174–92.
Kelley, M. R., and Helper, S. (1999), 'Firm size and capabilities, regional agglomeration, and the adoption of new technology', *Economics of Innovation and New Technology*, 8: 79–104.
Kirman, A. P. (1992), 'Variety: the coexistence of techniques', *Revue d' Économie Industrielle*, 59: 62–74.
Klepper, S. (1997), 'Industry life cycles', *Industrial and Corporate Change*, 6: 145–83.
Kline, S. J., and Rosenberg, N. (1986), 'An overview of innovation', in R. Landau and N. Rosenberg (eds.), *The Positive Sum Strategy*, Washington: National Academy Press.
Krugman, P. (1991*a*), *Geography and Trade*, Cambridge, Mass.: MIT Press.
—— (1991*b*), 'History versus expectations', *Quarterly Journal of Economics*, 106: 651–67.
—— (1996), *The Self-Organizing Economy*, Oxford: Blackwell.
Lamberton, D. (ed.) (1971), *Economics of Information and Knowledge*, Harmondsworth: Penguin.
—— (ed.) (1996), *The Economics of Information and Communication*, Cheltenham: Edward Elgar.
Langlois, R. N. (1992), 'External economies and economic progress: the case of the microcomputer industry', *Business History Review*, 66: 1–50.
—— and Robertson, P. L. (1995), *Firms, Markets, and Economic Change*, London: Routledge.
Latour, B. (1987), *Science in Action: How to Follow Scientists and Engineers in Society*, Cambridge, Mass.: Harvard University Press.
Layard, R., Nickell, S., and Jackman, R. (1994), *The Unemployment Crisis*, Oxford: Oxford University Press.
Lazonick, W. (1990), *Competitive Advantage on the Shop Floor*, Cambridge, Mass.: Harvard University Press.
—— (1991), *Business Organization and the Myth of the Market Economy*, Cambridge: Cambridge University Press.

Leoncini, R. (1998), 'The nature of long-run technological change: innovation, evolution, and technological systems', *Research Policy*, 27: 75–93.

Loasby, B. J. (1998), 'The organization of capabilities', *Journal of Economic Behaviour and Organization*, 35: 139–60.

—— (1999), *Knowledge Institutions and Evolution in Economics*, London: Routledge.

Lucas, R. (1988), 'On the mechanism of economic development', *Journal of Monetary Economics*, 22: 3–42.

Lundvall, B. A. (1985), *Product Innovation and User-Producer Interaction*, Aalborg: Aalborg University Press.

—— and Johnson, B. (1994), 'The learning economy', *Journal of Industry Studies*, 2: 23–42.

Machlup, F. (1962), *The Production and Distribution of Knowledge in the United States*, Princeton: Princeton University Press.

—— and Penrose, E. (1950), 'The patent controversy in the nineteenth century', *Journal of Economic History*, 10: 1–29.

Malerba, F. (1992), 'Learning by firms and incremental technical change', *Economic Journal*, 102: 845–59.

Manning, A. (1995), 'How do we know that real wages are too high?' *Quarterly Journal of Economics*, 110: 1111–25.

Mansfield, E. (1991), 'Academic research and industrial innovation', *Research Policy*, 20: 1–12.

Marshall, A. (1920/1961), *Principles of Economics*, London: Macmillan, 8th edn., 1961.

Mazzoleni, R., and Nelson, R. (1998), 'The benefits and costs of strong patent protection: a contribution to the current debate', *Research Policy*, 27: 273–84.

Metcalfe, J. S. (1995a), 'Technology systems and technology policy in historical perspective', *Cambridge Journal of Economics*, 19: 25–47.

—— (1995b), 'The economic foundation of technology policy: equilibrium and evolutionary perspectives', in Stoneman (1995).

—— (1997), *Evolutionary Economics and Creative Destruction*, London: Routledge.

—— and Miles, I. (eds.) (2000), *Innovation Systems in the Service Economy: Measurement and Case Study Analysis*, Dordrecht and Boston: Kluwer Academic Publishers.

Misa, T. J. (1995), 'Retrieving sociotechnical change from technological determinism', in Smith and Marx (1995).

Mokyr, J. (1990), *The Levels of Riches*, Oxford: Oxford University Press.

Mowery, D. C. (1983), 'The relationship between contractual and interfirm forms of industrial research in American manufacturing, 1900–1940', *Explorations in Economic History*, 20: 351–74.

Mowery, D. C. (1984), 'Firm structure, government policy, and the organization of industrial research: Great Britain and the United States, 1900–1950', *Business History Review*, 58: 504–31.

—— (1989), 'Collaborative ventures between U.S. and foreign manufacturing firms', *Research Policy*, 18: 19–32.

—— (1995), 'The boundaries of the U.S. firms in R&D', in N. R. Lamoreaux and D. M. G. Raff (eds.), *Coordination and Information*, Chicago: The University of Chicago Press for the National Bureau of Economic Research.

—— (ed.) (1996), *The International Computer Software Industry: A Comparative Study of Industry Evolution and Structure*, New York: Oxford University Press.

Mueller, D. C. (1987), *The Corporation: Growth, Diversification, and Mergers*, London: Harwood Academic Publishers.

—— and Tilton, J. (1969), 'Research and development costs as barriers to entry', *Canadian Journal of Economics*, 4: 570–9.

Musso, S. (1980), *Gli Operai di Turin: 1900–1920*, Milan: Feltrinelli.

Nelson, R. R. (1987), 'The role of knowledge in R&D efficiency', *Quarterly Journal of Economics*, 97: 453–70.

—— (ed.) (1993), *National Systems of Innovation*, Oxford: Oxford University Press.

—— (1998), 'The co-evolution of technology industrial structure, and supporting institutions', in G. Dosi, D. Teece, and J. Chytry (eds.), *Technology, Organization, and Competitiveness: Perspectives on Industrial and Corporate Change*, Oxford: Oxford University Press.

—— and Phelps, E. S. (1966), 'Investment in humans, technological diffusion, and economic growth', *American Economic Review*, 56: 69–75.

Noble, D. (1977), *America by Design: Science and Technology and the Rise of Corporate Capitalism*, New York: Oxford University Press.

Nordhaus, W. D. (1969), *Invention, Growth, and Welfare: A Theoretical Treatment of Technological Change*, Cambridge, Mass.: MIT Press.

Olson, M. (1965), *The Logic of Collective Action*, Cambridge, Mass.: Harvard University Press.

Ordover, J. A. (1991), 'A patent system for both diffusion and exclusion', *Journal of Economic Perspectives*, 5: 43–60.

Ostrom, E. (1990), *Governing the Commons: The Evolution of Institutions for Collective Action*, Cambridge: Cambridge University Press.

Paci, R., and Usai, S. (2000), 'Technological enclaves and industrial districts: an analysis of the regional distribution of innovative activity in Europe', *Regional Studies*, 34: 97–114.

Patel, P. (1995), 'The localised production of global technology', *Cambridge Journal of Economics*, 19: 141–53.

Pavitt, K. (1987), 'What we know about the strategic management of technology', *California Management Review*, 32: 17–26.
—— (1991), 'What makes basic research economically useful?' *Research Policy*, 20: 109–19.
—— (1999), *Technology Management and Systems of Innovation*, Cheltenham: Edward Elgar.
Penrose E. (1959/1980), *The Theory of the Growth of the Firm*, 2nd edn., Oxford: Basil Blackwell, 1980; 1st pub., 1959.
Perrin, J. (1991), 'Analyse des systèmes techniques', in Boyer *et al.* (1991).
Petit, P. (1988), *La Croissance Tertiaire*, Paris: Economica.
—— (1995), 'Employment and technological change', in Stoneman (1995).
Phelps, E. S. (1966), 'Models of technical progress and the golden rules of research', *Review of Economics and Statistics*, 48: 133–45.
Preissl, B. (1995), 'Strategic use of communication technology: diffusion processes in networks and environments', *Information Economics and Policy*, 7: 75–100.
Prendergast, C. (1999), 'The provision of incentives in firms', *Journal of Economic Literature*, 37: 7–63.
Quéré, M. (1994), 'Basic research inside the firm: lessons from an in-depth case study', *Research Policy*, 23: 413–24.
Reich, L. S. (1985), *The Making of American Industrial Research*, Cambridge: Cambridge University Press.
Richardson, G. B. (1960), *Information and Investment*, Oxford: Oxford University Press.
—— (1972), 'The organisation of industry', *Economic Journal*, 82: 883–96.
—— (1998), *The Economics of Imperfect Knowledge*, Aldershot: Edward Elgar.
Rizzello, S. (1999), *The Economics of the Mind*, Aldershot: Edward Elgar.
Robertson, P., and Langlois, R. N. (1995), 'Innovation networks and vertical integration', *Research Policy*, 24: 543–62.
Romer, P. M. (1986), 'Increasing returns and long-run economic growth', *Journal of Political Economy*, 94: 1002–37.
—— (1990), 'Endogenous technological change', *Journal of Political Economy*, 98: S71–102.
—— (1994), 'The origins of endogenous growth', *Journal of Economic Perspectives*, 8: 3–22.
Rosenberg, N. (1976), *Perspectives on Technology*, Cambridge: Cambridge University Press.
—— (1982), *Inside the Black Box: Technology and Economics*, Cambridge: Cambridge University Press.
—— and Nelson, R. (1992), 'American universities and technical advance in industry', *Research Policy*, 23: 323–48.

Russo, M. (1985), 'Technical change and the industrial district: the role of interfirm relations in the growth and transformation of the ceramic tile production in Italy', *Research Policy*, 14: 329–43.

—— (1996), *Cambiamento Tecnico e Relazioni tra Imprese*, Turin: Rosenberg & Sellier.

—— (2000), 'Complementary innovations and generative relationships: an ethnographic study', *Economics of Innovation and New Technology*, 9: 517–58.

Salais, R., and Storper, M. (1992), 'The four worlds of contemporary industry', *Cambridge Journal of Economics*, 16: 169–93.

Salop, S. C. (1979), 'A model of the natural rate of unemployment', *American Economic Review*, 69: 117–25.

Salter, W. E. G. (1966), *Productivity and Technical Change*, Cambridge: Cambridge University Press.

Samuelson, P. A. (1954), 'The pure theory of public expenditure', *Review of Economics and Statistics*, 36: 387–9.

—— (1955), 'Diagrammatic exposition of a pure theory of public expenditure', *Review of Economics and Statistics*, 37: 350–6.

Saxenian, A. (1994), *Regional Advantage: Culture and Competition in Silicon Valley and Route 128*, Cambridge, Mass.: Harvard University Press.

Scherer, F. M. (1984), *Innovation and Growth: Schumpeterian Perspectives*, Cambridge, Mass.: MIT Press.

—— (1999), *New Perspectives on Economic Growth and Technological Innovation*, Washington: Brookings Institution Press.

Schumpeter, J. A. (1934), *The Theory of Economic Development*, Cambridge, Mass.: Harvard University Press.

Scitowsky, T. (1954), 'Two concepts of external economies', *Journal of Political Economy*, 62: 143–51.

Schmookler, J. (1966), *Invention and Economic Growth*, Cambridge, Mass.: Harvard University Press.

Scotchmer, S. (1991), 'Standing on the shoulders of giants: cumulative research and the patent law', *Journal of Economic Perspectives*, 5: 29–41.

Shapiro, C., and Stiglitz, J. E. (1984), 'Equilibrium unemployment as a worker discipline device', *American Economic Review*, 74: 433–44.

—— and Varian, H. (1999), *Information Rules: A Strategic Guide to the Network Economy*, Boston: Harvard Business School Press.

Simon, H. A. (1958/1982), 'The role of expectations in an adaptive or behavioristic model', in M. J. Bowman (ed.), *Expectations, Uncertainty, and Business Behavior*, New York: Social Science Research Council, 1958. Reprinted in *Models of Bounded Rationality: Behavioral Economics and Business Organization*, Cambridge, Mass.: MIT Press, 1982.

—— (1962/1969), 'The architecture of complexity', *Proceedings of the American Philosophical Society*, 106 (1962): 467–82. Reprinted in *The Sciences of Artificial*, Cambridge, Mass.: MIT Press, 1969.

—— (1985), 'What do we know about the creative process?' in R. L. Kuhn (ed.), *Frontiers in Creative and Innovative Management*, Cambridge, Mass.: Ballinger.

Smith, M. R., and Marx, L. (eds.) (1995), *Does Technology Drive History? The Dilemma of Technological Determinism*, Cambridge, Mass.: MIT Press.

Solow, R. M. (1979), 'Another possible source of wage stickiness', *Journal of Macroeconomics*, 1: 79–82.

—— (1990), *The Labor Market as a Social Institution*, Oxford: Blackwell.

Stephan, P. E. (1996), 'The economics of science', *Journal of Economic Literature*, 34: 1199–1235.

Stigler, G. J. (1939), 'Production and cost theory in the short term', *Journal of Political Economy*, 47: 305–27.

—— (1966), *The Theory of Price*, London and New York: Macmillan.

Stiglitz, J. E. (1987), 'Learning to learn: localized learning and technological progress', in Dasgupta and Stoneman (1987).

Stiglitz, J. E. (1994), *Whither Socialism?* Cambridge, Mass.: MIT Press.

Stoneman, P. (ed.) (1995), *Handbook of the Economics of Innovation and Technological Change*, Oxford: Basil Blackwell.

Storper, M. (1995), 'The resurgence of regional economics, ten years later: the region as a nexus of untraded interdependencies', *European Urban and Regional Studies*, 2: 191–221.

—— (1996), 'Innovation as a collective action: conventions, products, and technologies', *Industrial and Corporate Change*, 5: 761–90.

—— Harrison, B. (1991), 'Flexibility, hierarchy, and regional development: the changing structure of industrial production systems and their forms of governance in the 1990s', *Research Policy*, 20: 407–22.

Sundbo, J. (1998), *The Organization of Innovation in Services*, Roskilde: Roskilde University Press.

Sutton, J. (1991), *Sunk Costs and Market Structure*, Cambridge, Mass.: MIT Press.

—— (1998), *Technology and Market Structure: Theory and Structure*, Cambridge, Mass.: MIT Press.

Swann, P. (ed.) (1994), *New Technologies and the Firm: Innovation and Competition*, London: Routledge.

—— Prevezer, M., and Stout, D. (eds.) (1998), *The Dynamics of Industrial Clustering*, Oxford: Oxford University Press.

Taylor, A. M. R., and Dixon, H. D. (1997), 'Controversy: on modelling the long run in applied econometrics', *Economic Journal*, 107: 165–8.

Teece, D. (1996), 'Firm organization, industrial structure, and technological innovation', *Journal of Economic Behavior and Organization*, 31: 193–224.

Teubal, M., Foray, D., Justman, M., and Zuscovitch, E. (eds.) (1996), *Technological Infrastructure Policy: an International Perspective*, Boston: Kluwer Academic Publishers.

Thurow, L. C. (1999), *Building Wealth*, New York: HarperCollins.

Utterback, J. M. (1994), *Mastering the Dynamics of Innovation*, Boston: Harvard Business School Press.

Varga, A. (1998), *University Research and Regional Innovation: A Spatial Econometric Analysis of Academic Technology Transfer*, Boston and Dordrecht: Kluwer Academic Publishers.

Volpato, G. (1996), *Il Caso Fiat: Una Strategia di Riorganizzazione e Rilancio*, Milan: ISEDI.

Von Hippel, E. (1988), *The Sources of Innovation*, London: Oxford University Press.

Weiss, A. (1980), 'Job queues and lay offs in labor markets with flexible wages', *Journal of Political Economy*, 88: 526–38.

Williamson, O. E. (1975), *Markets and Hierarchies: Analysis and Antitrust Implications*, New York: The Free Press.

—— (1985), *The Economic Institutions of Capitalism*, New York: The Free Press.

—— (1993), 'The logic of economic organization', in O. E. Williamson and S. G. Winter (eds.), *The Nature of the Firm*, Oxford: Oxford University Press.

Winter, S. G. (1987), 'Knowledge and competence as strategic assets', in D. J. Teece (ed.), *The Competitive Challenge*, Cambridge, Mass.: Ballinger.

Index

absorption costs 69–71
academic entrepreneurship 99–100
  see also universities
acquisitions 102, 149, 159, 190
adaptation 7
  and innovation 14–18
Adler, P. 58
ADSL 136
advanced local financial markets 93–4
agglomeration 74–5
  see also clusters; technological districts
Akerlof, G. A. 56, 57
Alfa Romeo 149
allocation analysis 72
Amendola, M. 142
Antonelli, C. 18, 47, 49, 60, 65, 102, 128, 141, 147, 187, 192
Aoki, M. 59
appropriability 77
Arora, A. 102, 186
Arrow, K. J. 36, 41, 44–5, 56, 79, 86, 183
Arthur, B. 118, 126
Atkinson, A. B. 2
ATM 134
Autobianchi 149

barter-based transactions 73
Bassignana, P. L. 158
Bessant, J. 58
BISDN 134
Bologna 92
broadband fibre optics 128, 136, 137, 139
business community-university interactions 99–100
business strategies 164–5

Cambridge 112
capital 29
  adoption of new capital goods 21
  stocks 19, 30, 31
  venture capital 94
Carlsson, B. 127
Castells, M. 102
centrifugal technologies 130, 132–3, 135, 137
centripetal technologies 130, 134–6, 137
Chandler, A. 147, 148, 169, 179, 183
channels, communication 83, 88, 89, 108
choices, technological 31–6
Clark, G. L. 3
Clark, K. B. 139
clusters, technological 9, 111–45, 201

industrial dynamics and new clusters 127–40
innovation policy and 210–11
irreversibility, innovation and 30, 43
key factors in system dynamics 113–19
system dynamics 119–27
codified knowledge 47–8, 97, 158–60, 162, 179–80
collective knowledge 9, 10–11, 69–80, 114, 116, 204–5
  externalities and transaction costs 75–80
  local externalities, absorption costs and 69–71
  technological districts 106–7, 109
  technological knowledge as 71–5
  tragedies of the commons 109
commons:
  innovation commons 6
  technological knowledge as 71–5
  tragedies of the commons 76–7, 80, 109
communication 8–9, 202–3
  channels 83, 88, 89, 108
  costs 82, 84–6, 88, 89–90
  generation and exchange of knowledge 45–6
  local communication infrastructure 102–3
  technological *see* technological communication
communication technologies 60, 102, 107, 116–19, 205–6
  industrial dynamics and new technological clusters 127–40
  and knowledge production 192–3
compatibility 30, 113–15
compensation schemes 206–8
competence 47, 48, 66
  internal 147
complementarities 91–2
  external and internal 35–6, 42
  local and global knowledge complementarity 50–1
  product and process 33, 42, 115
  synchronic and diachronic 35–6, 38, 41–2, 114–15, 196–7
  technological *see* technological complementarities
  technological clusters 112, 113–16, 121–2, 124, 141
complex systems 6, 50, 78, 169
computers 129
concentration 2–3, 5, 197–8
  regional 2–3, 5, 101, 197, 209–10

Index

concentration (cont.):
  technological 3, 5, 197
connectivity 91, 208
content providers 136
context-based innovation policy 209
convergence, technological 113, 123–4
cooperation
  industrial 186
  technological 104, 116–17
  in technological districts 103–5
core–periphery interaction 168, 170–1
corporate growth 147–9
  see also Fiat
corporations 104, 206–7
  see also large firms
costs:
  absorption costs 69–71
  communication costs 82, 84–6, 88, 89–90
  dynamic efficiency wages and 63
  incremental 118
  innovation costs 25–7, 50, 88
  R&D costs 123, 125–6
  reduction and innovation effort 33, 34
  short-term and long-term 20–1
  sunk costs 18, 19
  transaction costs 75–80, 203
craft organization of knowledge production 182–3, 193
creative effort 60–1
creative reactivity 7, 15–18, 198
creativity, employee 57–61
credit-rationing 93–4
cross-entries 135
cumulability 49, 73
  sequential 164
  systemic 49–50

data communication 131–4
David, P. A. 5, 53–4, 105, 106, 111, 115, 117, 169
decision-making 36–7
dedicated networks 132
demand 22–3
  labour demand 64
  network externalities 118
  segmentation 131–2
  technological clusters 115, 119–20
density, economies of 118
diachronic complementarities 35, 41–2, 114–15, 196–7
diffusion 21, 195–6
digital television 136
direct network complementarities 114
diversification 118–19, 121, 172
divestiture 133, 135
Dixit, A. 18, 115
dualism, new 208
dynamic efficiency wages 55–68, 189

Fiat 160–2
  and learning 56–61, 66
  and rates of generation of localized knowledge 61–5

economies of density 118
economies of growth 147–8
economies of scope 35, 38, 169, 179, 207
  endogenous 116, 121
education 100–1
efficiency 126
efficiency wages see dynamic efficiency wages
effort 60–1
electronic commerce 136
Emilia 152, 154, 155
employee involvement 56, 57–61, 66–7, 161–2, 188–9
employment 174
employment security 161
endogenous complementarities 31–6
endogenous economies of scope 116, 121
endogenous growth 36–40
endogenous technological change 165–71
endogenous technological externalities 116
endowments 28, 29, 30
entropy 30
entry 30, 98, 137–8
  cross-entries 135
  innovative 121–2, 141–2
  specialized 119, 121–2, 138
exaptation 17, 198
excludability 77
expectations 16–17, 36–7, 38–9
external complementarities 36, 42
external increasing returns 51–2, 70
external irreversible production factors 32
external knowledge 47–8, 97, 158–60, 179–80
externalities 35, 179, 203
  collective knowledge and transaction costs 75–80
  and increasing returns 105–6
  local see local externalities
  technological 38, 81–2, 83, 116, 203
Extranets 133–4

feedback, positive 52–3, 105–6, 201
Ferrari 149
Fiat 9–10, 146–80, 206–7
  accumulation of localized knowledge 153–65
  growth and technological change 149–77
  patent data and rate and direction of endogenous technological change 165–71
  statistical analysis 172–7
fibre optics 128, 136, 137, 139
financial markets, advanced local 93–4
firms 11

Index

incumbents 118–19, 121, 137
large 97–8, 101, 104, 132
learning firms 60, 64–5
new 98
  see also start-ups
number engaged in innovation 80
number in a technological district 89–91
performance and market selection of new technologies 142–4
small 97, 98, 101
fixed-input-intensive technologies 29
flexible-input-intensive technologies 29
foreign technology 159–60
Fransmann, M. 139
fungibility 30

Gaffard, J. L. 142
gales of innnovation 3–4, 111
Gambardella, A. 102, 186
gateway technologies 117
general-purpose technologies 3
generalized indivisibility 78
generic search 33, 34
generic technological knowledge 4
geodesic networks 84, 89–90
Geuna, A. 99, 187
Giacosa, D. 158, 160, 164
global knowledge externalities 78
Granger causality test 174
Griliches, Z. 61, 77, 174
growth:
  corporate 147–9
    see also Fiat
  economies of 147–8
  endogenous 36–40
  new growth theory 4–5

Henderson, R. 139
high-tech start-ups 97–8, 101, 190
Hirschleifer, J. 49
history 16–17
horizontal indivisibility 49

increasing returns 4, 16, 41, 91, 197–8, 201
  collective technological knowledge and externalities 75–80
  external 51–2, 70
  externalities and positive feedback 105–6
  Fiat 176–7
  localized technological knowledge 51–4
incumbents 137
  diversification and vertical integration 118–19, 121
indirect network complementarities 114
indivisibility 49–50
  modular and generalized 78
induced localized technological change 22–31, 36
industrial clubs 104
industrial concentration 3
industrial districts 98, 103, 203–4

industrial dynamics 9, 112, 141–2, 197
  and new technological clusters 127–40
industrial policy 143–4, 208–11
industrial structures, local 96–8
inflation 37–8
information:
  asymmetries in financial markets 93–4
  and knowledge 44–5, 46–7, 70
innovation 13–43, 118, 181–2, 195–6
  adaptation and 14–18
  commons 6
  costs 25–7, 50, 88
  efforts and cost reduction 33, 34
  gales of 3–4, 111
  knowledge, information and 46–7
  local innovation systems 65, 67–8, 74–5, 91–2, 99–102, 200
  positive feedback 105–6
  regional concentration 2–3
  technological concentration 3
innovation policy 10, 110, 144–5, 208–11
innovative entry 121–2, 141–2
intellectual enclosures 204
intellectual property rights 45, 76–7, 184, 187
intelligent antennas 136
intelligent networks 135–6
interactive television 128
inter-component linkages 50
interdependence 113–14
interest rates 37–8
intermediary markets, local 94–6
internal complementarities 35–6, 42
internal irreversible production factors 32
internal knowledge 47–8, 158–60, 179–80
internal labour markets 59–60, 162
Internet 128, 129, 136–7, 139
Internet Protocol 137
interoperability 113–15, 117
interregional links 98–9
intersectoral communication 96
intra-component linkages 50
Intranets 132–3
intrasectoral communication 96–7
involvement, employee 56, 57–61, 66–7, 161–2, 188–9
irreversibility 7, 13–43, 118, 198
  Fiat 153, 172
  increasing returns, externalities and positive feedback 105–6
  inducement of localized technological change 22–31, 36
  out-of-equilibrium and endogenous growth 36–40
  and technological change 18–22
  technological choices and endogenous complementarities 31–6
ISDN 134, 135–6
Italian car industry 149–52, 154–5
  see also Fiat

just-in-time management 133

Kaldor, N. 148
Kirman, A. P. 34
Klepper, S. 142
Kline, S. J. 46
knowledge:
 codified 47–8, 97, 158–60, 162, 179–80
 collective *see* collective knowledge
 external 47–8, 97, 158–60, 179–80
 internal 47–8, 158–60, 179–80
 localized technological knowledge *see* localized technological knowledge
 near-decomposability of 6, 49, 50–1, 72–3
 tacit *see* tacit knowledge
 technological *see* technological knowledge
knowledge externalities 123
knowledge infrastructure 99–102
knowledge-intensive business services 186–7, 189
 academic supply of 99–100
 dynamic efficiency wages 55, 58, 65, 66–7
 technological communication 85, 95, 99–100, 110
knowledge production 10, 54, 181–94, 206–7
 craft organization 182–3, 193
 emergence of localized knowledge model 185–93, 194
 R&D model 183–5, 193–4
 technological communication 86–7, 91–2
knowledge technological externalities 77–8
Krugman, P. 17, 36

labour demand 64
labour markets:
 internal 59–60, 162
 local conditions 93
 new dualism 208
labour productivity 56
Lancia 149
Langlois, R. N. 119
large firms 132
 technological districts 97–8, 101, 104
 see also corporations
Latour, B. 195
Lazio 152, 154, 155
learning 9, 44–5, 46, 188
 dynamic efficiency wages and 56–61, 66
 Matthew effect 53–4
learning firms 60, 64–5
liberalization 133, 135
licences 101–2
Liguria 152, 154, 155
local communication infrastructure 102–3
local externalities 198–9
 absorption costs and collective knowledge 69–71
 collective knowledge and transaction costs 78–80

local financial markets, advanced 93–4
local increasing returns 51–2
local industrial structures 96–8
local innovation systems 65, 67–8, 74–5, 91–2, 200
 knowledge infrastructure 99–102
local intermediary markets 94–6
local knowledge externalities 43, 78, 82–3
local labour markets 93
localization 103–5
localized search 33, 34
localized technological change 1–2, 38–9, 200–1
 and corporate growth 147–9, 178–80
 Fiat 172–7
 inducement of 22–31, 36
localized technological knowledge 1–2, 44–54, 199–200
 accumulation by Fiat 153–65
 as a collective activity 69–80
 and corporate growth 147–9
 dynamic efficiency wages and 57–9, 65–7
 rates of generation 61–5
 model of knowledge production 185–93, 194
location 19, 39–40
Lombardy 152, 154, 155
long-term cost theories 20–1

Malerba, F. 46
market selection of technologies 142–4
Marshall, A. 3, 18
Marshallian industrial districts 98, 103, 203–4
Matthew effect 53–4, 106
mergers and acquisitions 102, 149, 159, 190
Metcalfe, J. S. 4, 46
metropolitan areas 102–3
microelectronics 129
mobile telephony 128, 131, 134–6, 137
Modena 92
modular indivisibility 78
modularity 6, 49, 50–1, 72–3
Mowery, D. C. 135, 183
multi-purpose networks 137
multi-regional firms 98–9

near-decomposability of knowledge 6, 49, 50–1, 72–3
Nelson, R. R. 28, 92, 187
network externalities 114, 118
networks, communication 83–5, 89–91, 107–8, 110
 telecommunications 127–40
new dualism 208
new firms 98
 see also start-ups
new growth theory 4–5
niche markets 119, 121–2, 138

Index

oligarchic networks 85, 90
online communication systems 192–3
Ordover, J. A. 187
organized networks 85, 89, 90
out-of-equilibrium conditions 14–15
  irreversibility and endogenous growth 36–40
output growth 148, 174–7
outsourcing 97
  technological outsourcing 186–7

participation, employee 56, 57–61, 66–7, 161–2, 188–9
Patel, P. 99
patents 101–2
  Fiat 156, 157, 165–72, 173–7
path dependence 41, 43, 201–2
Pavitt, K. 190
pecuniary complementarities 114
Penrose, E. 147–8
periphery–core interaction 168, 170–1
personal computers 129
pharmaceutical industry 185
Politecnico of Turin 159
positive feedback 52–3, 105–6, 201
post-sale assistance 73
Preissl, B. 135
price-cap regulation 144
price discrimination 131
private consulting 100
private goods 79
privatization 133, 135
process complementarities 33, 42, 115
process innovations 17, 164, 167
  interaction with product innovations 168, 171–2
product complementarities 35, 42, 115
product innovations 17, 164–5, 167
  interaction with process innovations 168, 171–2
productivity
  labour 56
  total factor 34–5, 61–5
professional associations 104
profit equation 26–7, 124–5
profitability 142–3
property rights 76–7, 109
  intellectual 45, 76–7, 184, 187
proximity 111–12
  collective technological knowledge and transaction costs 75–80
public goods 71–5, 79
public subsidies 45

quasi-losses 23–4, 27, 28, 40
quasi-rents 18

rationing 77
  credit-rationing 93–4
reactivity, creative 7, 15–18, 198

receptivity 91, 208
regional concentration 2–3, 5, 101, 197, 209–10
regional space 1–2
regional specialization 188
regions 7–8, 199, 200
  interregional links 98–9
relative prices, changes in 22–3
rent technological externalities 77–8
research and development (R&D) 181–2, 190
  costs 123, 125–6
  Fiat R&D unit 160
  model of knowledge production 183–5, 193–4
  regional concentration of laboratories 101
Richardson, G. B. 103, 104, 105
rivalry, technological 118
Robertson, P. L. 119
Romer, P. M. 4
Rosenberg, N. 46, 187
Route 128 92, 112
Rush, H. 58

Salter, W. E. G. 21, 31
Samuelson, P. A. 72
satellites 128–9
Scherer, F. M. 4–5, 77, 94
Schumpeter, J. A. 3–4, 111
scientific knowledge 183–5
Scitowsky, T. 77
scope, economies of *see* economies of scope
segmental technologies 130, 132
selective innovation policy 10, 210–11
sequential cumulability 164
short-term cost theories 20–1
Silicon Valley 92, 112
Simon, H. A. 6, 46, 50, 79, 199
small firms 97, 98, 101
social constructionism 195
Solow, R. M. 56
SPA 149
space technology 128–9, 131
specialization 130, 132, 188, 191
specialized entry 119, 121–2, 137
specific technological knowledge 4
spin-offs 190–1
standard effort 60
standards 117
Stankiewicz, R. 127
start-ups 99–100, 101, 102
  high-tech 97–8, 101, 190
Steinmueller, E. 117
Stephan, P. E. 49
Stigler, G. J. 20
Stiglitz, J. E. 2, 77, 109
structured networks 84–5, 89, 90
sub-clusters 139, 141
subsidies, public 45
subsystems 79

sunk costs 18, 19
superfixed production factors 18–21, 40–1
  endowment of 28, 29, 30
  see also irreversibility
supplier relationships 133
supply 115, 119–20
  network externalities 118
Sutton, J. 18
Swann, P. 98
switches 85
synchronic complementarities 35, 41–2, 114–15, 196–7
system dynamics 11, 119–27, 139–44, 196–7, 205–6
  key factors in 113–19
  system unbalances at Fiat 168, 169
systemic cumulability 49–50

tacit knowledge 47–8, 59–60
  craft organization of knowledge production 182–3
  Fiat 158–60, 162, 179–80
  technological communication 95, 97
team work 57
technical space 1–2
technological change 4, 158, 200, 202
  direction of 36–7, 140
  irreversibility and 18–22
  localized *see* localized technological change
  rate of 29, 36–7, 39, 140
  rate and direction at Fiat 165–72
  system dynamics 115–16
  telecommunications industry 135
technological choices 31–6
technological clubs 192–3
technological clusters *see* clusters, technological
technological communication 65, 68, 81–110, 202–3
  economics of 81–6
  Fiat 162–4
  geometric exposition 86–92
  increasing returns, externalities and positive feedback 105–6
  innovation policy and 209–10
  and innovation within technological districts 93–105, 202–5
  in localized technological change 186, 192–3, 200–1
  regional innovation systems 74–5
technological complementarities 30–1, 41–2, 148–9, 196–7
  clusters and 114, 122, 124, 141
  and growth 179
  intentional search for 206–8
  irreversibility, out-of-equilibrium and endogenous growth 38
  irreversibility, technological choices and 31–6
technological concentration 3, 5, 197
technological convergence 113, 123–4

technological cooperation 104, 116–17
technological districts 43, 88–9, 92, 106–10, 188, 206–7
  increasing returns, externalities and positive feedback 105–6
  innovation policy 209–10
  technological communication and innovation within 93–105, 202–5
technological externalities 38, 81–2, 83, 116, 203
  see also externalities
technological innovation systems 74–5
technological knowledge:
  generic and specific 4
  localized *see* localized technological knowledge
  public goods vs collective commons 71–5
technological outsourcing 186–7
technological path dependence 43, 201–2
technological rivalry 118
technological strategies 103–5
  and business strategies at Fiat 164–5
  firms' intentional action 206–8
technological systems 41, 196–8
  system dynamics *see* system dynamics
technological variety 33–4
technology production function 61
telecommunications 127–40
telephone industry 130–1
  mobile telephony 128, 131, 134–6, 137
television 128, 136
Thurow, L. C. 77
total factor productivity 34–5, 61–5
Toulouse 92
tragedies of the commons 76–7, 80, 109
training 100–1
transaction costs 75–80, 203
triangular communication 95–6
Turin 10, 92, 146, 152, 172–3
Tuscany 152, 154, 155

UMTS 136
universities 70–1
  academic entrepreneurship and supply of knowledge-intensive business services 99–100
  knowledge production 184–5, 187–8
usage innovations 117
user–producer interactions 117, 133
Utterback, J. 171

variety, technological 33–4
Veneto 152, 154, 155
venture capital 94
vertical indivisibility 49
vertical integration 95, 118–19
Volpato, G. 159

wages *see* dynamic efficiency wages
Winograd, T. 58
World Wide Web 135

The manufacturer's authorised representative in the EU for product safety is Oxford University Press España S.A. of el Parque Empresarial San Fernando de Henares, Avenida de Castilla, 2 – 28830 Madrid (www.oup.es/en or product.safety@oup.com). OUP España S.A. also acts as importer into Spain of products made by the manufacturer.

www.ingramcontent.com/pod-product-compliance
Lightning Source LLC
LaVergne TN
LVHW041205250326
834689LV00001BA/10